Communications in Computer and Information Science   **656**

Commenced Publication in 2007
Founding and Former Series Editors:
Alfredo Cuzzocrea, Dominik Ślęzak, and Xiaokang Yang

More information about this series at http://www.springer.com/series/7899

Juan Antonio Lossio-Ventura
Hugo Alatrista-Salas (Eds.)

Information Management and Big Data

Second Annual International Symposium, SIMBig 2015
Cusco, Peru, September 2–4, 2015
and Third Annual International Symposium, SIMBig 2016
Cusco, Peru, September 1–3, 2016
Revised Selected Papers

 Springer

Editors
Juan Antonio Lossio-Ventura
University of Florida
Gainesville, FL
USA

Hugo Alatrista-Salas
Universidad del Pacífico
Jesús María, Lima
Peru

ISSN 1865-0929   ISSN 1865-0937   (electronic)
Communications in Computer and Information Science
ISBN 978-3-319-55208-8   ISBN 978-3-319-55209-5   (eBook)
DOI 10.1007/978-3-319-55209-5

Library of Congress Control Number: 2017933877

Printed on acid-free paper

This Springer imprint is published by Springer Nature
The registered company is Springer International Publishing AG
The registered company address is: Gewerbestrasse 11, 6330 Cham, Switzerland

Preface

The aim of the SIMBig symposium is to present the analysis of methods for extracting knowledge from large volumes of data through techniques of data science and artificial intelligence. This books comprises extended versions of the best papers presented at SIMBig 2015 and SIMBig 2016.

Big data is a popular term used to describe the exponential growth and availability of data, which could be structured and unstructured. Data science is a field seeking to extract knowledge or insights from large volumes of heterogeneous data (e.g., video, audio, text, image). Data science is a continuation of other fields such as data analysis, statistics, machine learning, and data mining similar to knowledge discovery in databases (KDD).

Big data has taken place over the past 20 years. For instance, social networks such as Facebook, Twitter, and LinkedIn generate masses of data, which are available to be accessed by other applications. Several domains, including biomedicine, life sciences, and scientific research, have been affected by big data[1]. Therefore, there is a need to understand and exploit these data. This process can be carried out thanks to "data science", which is based on methodologies of data mining, natural language processing, Semantic Web, statistics, etc. That allows us to gain new insight through data-driven research [1, 4]. A major problem hampering big data analytics development is the need to process several types of data, such as structured, numeric, and unstructured data (e.g., video, audio, text, image, etc.)[2].

Our Annual International Symposium on Information Management and Big Data, seeks to present the new methods of data science and related fields for analyzing and managing large volumes of data. The symposium attracts many of the main national and international players in the decision-making field who are involved in new technologies dedicated to handling large amounts of information.

The third edition, SIMBig 2016[3], was held in Cusco, Peru, during September 1–3, 2016. SIMBig 2016 has been indexed in DBLP[4] [3] and in the CEUR Workshop Proceedings[5]. The second edition, SIMBig 2015[6], was also held in Cusco, Peru, during September 2–4, 2015. SIMBig 2015 has been indexed in DBLP[7] [2] and in the CEUR Workshop Proceedings[8].

---

[1] By 2015 the average amount of data generated annually in hospitals is 665TB: https://datafloq.com/read/body-source-big-data-infographic/413.

[2] Today, 80% of data are unstructured such as images, video, and notes.

[3] http://simbig.org/SIMBig2016/.

[4] http://dblp2.uni-trier.de/db/conf/simbig/simbig2016.html.

[5] http://ceur-ws.org/Vol-1743/.

[6] http://simbig.org/SIMBig2015/.

[7] http://dblp2.uni-trier.de/db/conf/simbig/simbig2015.html.

[8] http://ceur-ws.org/Vol-1478/.

For this special proceedings volume, we accepted 11 long papers. These papers were selected from the SIMBig 2015 and SIMBig 2016 editions. We selected the four best papers from the 2015 edition, which had 32 submissions. Likewise, we selected the seven best papers of SIMBig 2016, from 42 submissions. Therefore, the general acceptance rate of this special proceedings volume was $11/(42 + 32) = 0.15$.

To share the new analysis methods for managing large volumes of data, we encouraged participation from researchers in all fields related to big data, data science, data mining, natural language processing, and the Semantic Web, but also multilingual text processing as well as biomedical NLP.

Topics of interest of SIMBig included: data science, big data, data mining, natural language processing, Bio-NLP, text mining, information retrieval, machine learning, Semantic Web, ontologies, Web mining, knowledge representation and linked open data, social networks, social Web, and Web science, information visualization, OLAP, data warehousing, business intelligence, spatiotemporal data, health care, agent-based systems, reasoning and logic, constraints, satisfiability, and search.

February 2017   Juan Antonio Lossio-Ventura
                                                              Hugo Alatrista-Salas

References

1. David W Embley and Stephen W Liddle, *Big data—conceptual modeling to the rescue*, Conceptual Modeling, ER'13, LNCS, Springer, 2013, pp. 1–8.
2. Juan Antonio Lossio-Ventura and Hugo Alatrista-Salas (eds.), *Proceedings of the 2nd annual international symposium on information management and big data - simbig 2015, cusco, Peru, September 2–4, 2015*, CEUR Workshop Proceedings, vol. 1478, CEUR-WS.org, 2015.
3. Juan Antonio Lossio-Ventura and Hugo Alatrista-Salas (eds.), *Proceedings of the 3rd annual international symposium on information management and big data - simbig 2016, cusco, Peru, September 1–3, 2016*, CEUR Workshop Proceedings, vol. 1743, CEUR-WS.org, 2016.
4. Sam Madden, *From databases to big data*, vol. 16, IEEE Educational Activities Department, Piscataway, NJ, USA, May 2012, pp. 4–6.

Organizing Committee

General Organizers

Juan Antonio   University of Florida, USA
  Lossio-Ventura
Hugo Alatrista-Salas   Universidad del Pacífico, Peru

Local Organizers

Cristhian Ganvini Valcarcel   Universidad Andina del Cusco, Peru
Armando Fermin Perez   Universidad Nacional Mayor de San Marcos, Peru

Program Committee

Nathalie Abadie   French National Mapping Agency, COGIT, France
Elie Abi-Lahoud   University College Cork, Cork, Ireland
Salah AitMokhtar   Xerox Research Centre Europa, France
Sophia Ananiadou   NaCTeM - University of Manchester, UK
Marcelo Arenas   Pontificia Universidad Catolica de Chile, Chile
Jérôme Azé   LIRMM - University of Montpellier, France
Riza Batista-Navarro   NaCTeM - University of Manchester, UK
Pablo Barceló   Universidad de Chile, Chile
Nicolas Béchet   IRISA - University of Bretagne Sud, France
Cesar A. Beltrán Castañón   GRPIAA - Pontifical Catholic University of Peru, Peru
Lilia Berrahou   LIRMM - University of Montpellier, France
Jiang Bian   College of Medicine, University of Florida, USA
Albert Bifet   Télécom ParisTech, France
Sandra Bringay   LIRMM - Paul Valéry University, France
Mohamed R. Bouadjenek   University of Melbourne, Australia
Thierry Charnois   GREYC Université Paris 13, France
Oscar Corcho   Ontology Engineering Group - Polytechnic University
                                 of Madrid, Spain
Bruno Crémilleux   GREYC-CNRS, Université de Caen Normandie,
                                 France
Fabio Crestani   University of Lugano, Switzerland
Gabriela Csurka   Xerox Research Centre Europa, France
Martín Ariel Domínguez   Universidad Nacional de Córdoba, Argentina
Paula Estrella   Universidad Nacional de Córdoba, Argentina
Frédéric Flouvat   PPME Lab - University of New Caledonia,
                                 New Caledonia

Philippe Fournier-Viger	Harbin Institute of Technology Shenzhen Graduate School, China
André Freitas	University of Passau, Germany
Mauro Gaio	LIUPPA - University of Pau, France
Natalia Grabar	CNRS - University of Lille 3, France
Adrien Guille	Université Lumière Lyon 2, France
Thomas Guyet	IRISA/LACODAM - Agrocampus Ouest, France
Hakim Hacid	Zayed University, United Arab Emirates
Phan Nhat Hai	Oregon State University, USA
Sébastien Harispe	Ecole des Mines d'Alès, France
Dino Ienco	Irstea, France
Diana Inkpen	University of Ottawa, Canada
Clement Jonquet	LIRMM - University of Montpellier, France
Alípio Jorge	Universidade do Porto, Portugal
Eric Kergosien	GERiCO Lab - University of Lille 3, France
Georgios Kontonatsios	NaCTeM - University of Manchester, UK
Yannis Korkontzelos	NaCTeM - University of Manchester, UK
Christian Libaque Saenz	Universidad del Pacífico, Peru
Cédric López	VISEO - Research and Development Unit, France
Franco M. Luque	Universidad Nacional de Córdoba, Argentina
Florent Masseglia	Inria - Zenith Team, France
Peter Mika	Yahoo! Research Labs - Barcelone, Spain
André Miralles	SISO Team, Irstea, France
François Modave	College of Medicine: University of Florida, USA
Giovanni Montana	Imperial College London, UK
Nhung Nguyen	University of Manchester, UK
Jordi Nin	BBVA Data & Analytics and Universidad de Barcelona, Spain
Miguel Nuñez del Prado Cortez	Universidad del Pacífico, Peru
Maciej Ogrodniczuk	Institute of Computer Science, Polish Academy of Sciences, Poland
Thomas Opitz	Biostatistics and Spatial Processes - Inra, France
José Manuel Perea-Ortega	University of Extremadura, Spain
Yoann Pitarch	IRIT - Toulouse, France
Marc Plantevit	LIRIS-CNRS, Université Claude Bernard Lyon 1, France
Pascal Poncelet	LIRMM - University of Montpellier, France
Perfecto Quintero-Flores	Instituto Tecnológico de Apizaco, Mexico
Julien Rabatel	Catholic University of Leuven, Belgium
José Luis Redondo García	Ontology Engineering Group - Polytechnic University of Madrid, Spain
Mathieu Roche	Cirad - TETIS - LIRMM, France
Nancy Rodriguez	LIRMM - University of Montpellier, France
Avid Roman-Gonzales	Universidad Nacional Cayetano Heredia, Peru
Fatiha Saïs	Paris-Sud 11 University, France

Organizing Institutions and Sponsors

Universidad Andina del Cusco, Cusco, Peru
University of Florida, Florida, USA
Universidad del Pacífico, Lima, Peru
Pontificia Universidad Católica del Perú, Lima, Peru
Laboratoire de Informatique, Robotique et Micro-électronique de Montpellier, Montpellier, France
Université de Montpellier, Montpellier, France

Contents

Sense-Level Semantic Clustering of Hashtags

Ali Javed and Byung Suk Lee[⊠]

Department of Computer Science, University of Vermont, Burlington, VT, USA
{ajaved,bslee}@uvm.edu

Abstract. We enhance the accuracy of the currently available semantic hashtag clustering method, which leverages hashtag semantics extracted from dictionaries such as Wordnet and Wikipedia. While immune to the uncontrolled and often sparse usage of hashtags, the current method distinguishes hashtag semantics only at the word-level. Unfortunately, a word can have multiple senses representing the exact semantics of a word, and, therefore, word-level semantic clustering fails to disambiguate the true sense-level semantics of hashtags and, as a result, may generate incorrect clusters. This paper shows how this problem can be overcome through sense-level clustering and demonstrates its impacts on clustering behavior and accuracy.

Keywords: Sense-level · Semantic clustering · Hashtag

1 Introduction

Hashtag clustering has emerged as an interesting and important topic of study in online social media, which is arguably the best source of timely information. On Twitter alone, an average of 6,000 micro-messages are posted per second [14]. Thus, social media analysts use clusters of hashtags as the basis for more complex tasks on tweets [6] such as retrieving relevant tweets [6,7], tweet ranking, sentiment analysis [17], data visualization [1], semantic information retrieval [11], and user characterization. Therefore, the accuracy of hashtag clustering is important to the quality of the resulting information in those tasks.

The popular approach to hashtag clustering has been to leverage the tweet texts accompanying hashtags [1,2,6,8,11–13] by identifying their "contextual" semantics [9]. There are two prominent problems with this approach, however. First, tweet texts are limited to only 140 characters in length and, moreover, a majority of hashtags are not used frequently enough to find sizable tweet texts accompanying them, thus causing a sparsity problem. Second, tweet texts are open-ended, with no control over their contents at all, and therefore often exhibit poor linguistic quality. (According to 2009 Pear Analytics Twitter Study, 40% of tweets are "pointless babble" [4].) These problems make text-based techniques ineffective for hashtag clustering. Hence, methods that utilize other means to identifying semantics of hashtags are needed.

In this regard, the focus of this paper is on leveraging *dictionary metadata* to identify the semantics of hashtags. We adopt the pioneering work done by

© Springer International Publishing AG 2017
J.A. Lossio-Ventura and H. Alatrista-Salas (Eds.): SIMBig 2015/2016, CCIS 656, pp. 1–16, 2017.
DOI: 10.1007/978-3-319-55209-5_1

Vicient and Moreno [15, 16]. Their approach identifies the "lexical" semantics of hashtags from external resources (e.g., Wordnet, Wikipedia) independent of the tweet messages themselves. To the best of our knowledge, their work is the only one that uses this metadata-based approach. This approach has the advantage of being immune to the sparsity and poor linguistic quality of tweet messages, and the results of their work demonstrate it.

On the other hand, their work has a major drawback, in that it makes clustering decisions at the *word* level while the correct decision can be made at the *sense* (or "concept") level. Obviously the correct use of metadata is critical to the performance of any metadata-based approach, and indeed clustering hashtags based on their word-level semantics has been shown to erroneously putting hashtags of different senses in the same cluster (more on this in Sect. 4).

In this paper, we devise a more accurate sense-level metadata-based semantic clustering algorithm. The critical area of improvement is in the construction of similarity matrix between pairs of hashtags, which then is input to a clustering algorithm. The immediate benefits are shown in the accuracy of resulting clusters, and we demonstrate it using a toy example. Experimental results using two gold standard tests showed gains of 26% (when hashtag semantics are not controlled) and 47% (when controlled), respectively, in terms of the weighted average pairwise maximum f-score (Eq. 5), where the weight is the size of a ground truth cluster. Despite the gain in the clustering accuracy, we were able to keep the runtime and space overheads for similarity matrix construction within a constant factor (e.g., 5 to 10) through a careful implementation scheme.

This paper contains more rigorous experiments than the authors' conference paper [3], which was invited into this CCIS Series with extended content.

The remainder of this paper is organized as follows. Section 2 provides some background knowledge. Section 3 describes the semantic hashtag clustering algorithm designed by Vicient and Moreno [16]. Section 4 discusses the proposed *sense*-level semantic enhancement to the clustering algorithm, and Sect. 5 presents its evaluation against the word-level semantic clustering. Section 6 presents other work related to the semantic hashtag clustering. Section 7 summarizes the paper and suggests future work.

2   Background

2.1   Wordnet – Synset Hierarchy and Similarity Measure

Wordnet groups English words into sets of synonyms called synsets. Synsets in Wordnet are interlinked by their semantics and lexical relationships, which results in a network of meaningful related words and concepts. The concepts are linked to each other using the semantic and lexical relationships mentioned. Given this network of relationships, we use the Wu-Palmer [18] similarity measure in order to stay consistent with the baseline algorithm by Vicient and Moreno [16]. Given concepts organized in a hierarchy, the Wu-Palmer similarity, $\text{sim}_{WP}(C_1, C_2)$, between two concepts C_1 and C_2 is defined as

$$\text{sim}_{WP}(C_1, C_2) = \frac{2 \cdot \text{depth}(\text{LCS}(C_1, C_2))}{\text{depth}(C_1) + \text{depth}(C_2)} \tag{1}$$

where $\text{LCS}(C_1, C_2)$ is the least common subsumer of C_1 and C_2 in the hierarchy. We use this Wordnet functionality to calculate the semantic similarity between hashtags, that is, by grounding hashtags to specific concepts (called "semantic grounding") and calculating the similarity between the concepts.

2.2 Wikipedia – Auxiliary Categories

Wikipedia is the most popular crowd-sourced encyclopedia. Not all hashtags can be grounded semantically using Wordnet because many of them are simply not legitimate terms found in Wordnet (e.g. #Honda). This situation is where Wikipedia can be used to look up those hashtags. Wikipedia provides auxiliary categories for each article. For example, when Wikipedia is queried for categories related to the page titled "Honda", it returns the auxiliary categories shown in Fig. 1.

Auxiliary categories can be thought of as categories the page belongs to. In this example, if we are unable to look up the word "Honda" on Wordnet, then, through the help of these auxiliary categories, we can relate the term to Japan, Automotive, Company, etc. There are several open source Wikipedia APIs available to achieve this purpose – for example, the Python library "wikipedia".

```
[Automotive companies of Japan',
 Companies based in Tokyo',
 Boat builders',
 Truck manufacturers',
 Vehicle manufacturing companies',
 ...
]
```

Fig. 1. Wikipedia auxilary categories for "Honda".

2.3 Hierarchical Clustering

Hierarchical clustering is a viable approach to cluster analysis, and is particularly suitable for the purpose of hashtag clustering in this paper. There are two strategies for hierarchical clustering – bottom-up (or agglomerative) and top-down (or divisive) – and bottom-up strategy is used in our work because it is conceptually simpler than top-down [5]. Several distance measures are available to provide linkage criteria for building up a hierarchy of clusters. Among them, single-linkage method and unweighted pair group method with arithmetic mean (UPGMA) are used most commonly. Single-linkage method calculates the distance between two clusters C_u and C_v as

$$d(C_u, C_v) = \min_{u_i \in C_u \wedge v_j \in C_v} \text{dist}(u_i, v_j) \tag{2}$$

and UPGMA calculates the distance as

$$d(C_u, C_v) = \sum_{u_i \in C_u, v_j \in C_v} \frac{d(u_i, v_j)}{|C_u| \times |C_v|} \tag{3}$$

where $|C_u|$ and $|C_v|$ denote the number of elements in clusters C_u and C_v, respectively.

To generate output clusters, we extract "flat clusters" from the hierarchy using the *distance* criterion, which, given a distance measure, forms flat clusters from the hierarchy when items in each cluster are no farther than a distance threshold.

3   Semantic Hashtag Clustering

The semantic clustering approach proposed by Vicient and Moreno [16] uses Wordnet and Wikipedia as the metadata for identifying the lexical semantics of a hashtag. Source codes of their algorithms were not available, and so we implemented the approach described in Vicient's PhD dissertation [15] to the best of our ability.

There are three major steps in their semantic clustering algorithm [16]: (a) semantic grounding, (b) similarity matrix construction, and (c) semantic clustering. Algorithm 1 summarizes the steps.

Input: list H of hashtags
Output: clusters
Stage 1 (Semantic grounding):
Step 1: For each hashtag $h \in H$ perform Step 1a.
    Step 1a: Look up h from Wordnet. If h is found then *append* the synset of h
    to a list (LC_h). Otherwise segment h into multiple words and drop the
    leftmost word and then try Step 1a again using the reduced h until either a
    match is found from Wordnet or no more word is left in h.
Step 2: For each $h \in H$ that has an empty list LC_h, look up h in Wikipedia. If
    an article matching h is found in Wikipedia, acquire the list of auxiliary
    categories for the article, extract main nouns from the auxiliary categories, and
    then, for each main noun extracted, go to Step 1a using the main noun as h.
Stage 2 (Similarity matrix construction): Discard any hashtag h that has
    an empty LC_h. Calculate the maximum pairwise similarity between each pair
    of lists LC_{h_i} and LC_{h_j} $(i \neq j)$ using any ontology-based similarity measure.
Stage3 (Clustering): Perform clustering on the distance matrix (1's
    complement of the similarity matrix) resulting from Stage 2.

Algorithm 1. Semantic hashtag clustering [16].

In the first stage (i.e., semantic grounding), each hashtag is looked up in Wordnet. If there is a direct match, that is, the hashtag is found in Wordnet, then it is added as a single candidate synset, and, accordingly, all the concepts (or senses) (see Sect. 2.1) belonging to the synset are saved in the form of a list of candidate concepts related to the hashtag. We call this list LC_h. If, on the other hand, the hashtag is not found in Wordnet, then the hashtag is split into multiple terms (using a word segmentation technique) and, then, the leftmost term is dropped sequentially until either a match is found in Wordnet or there is no more term left.

For each hashtag that was not found from Wordnet in Step 1 (i.e., of which the LC_h is empty), it is looked up in Wikipedia. If a match is found in Wikipedia, the auxiliary categories (see Sect. 2.2) of the article are acquired. Main nouns from the auxiliary categories are then looked up in Wordnet, and if a match is found, we save the concepts by appending them to the list LC_h; this step is repeated for each main noun.

In the second stage (i.e., similarity matrix construction), first, hashtags associated with an empty list of concepts are discarded; in other words, hashtags that did not match any Wordnet entry, either by themselves or by using word segmentation technique, and also had no entry found in Wikipedia are discarded. Then, using the remaining hashtags (each of whose LC_h contains at least one concept in it), semantic similarity is calculated between each pair of them. Any ontology-based measure can be used, and Wu-Palmer measure [18] (see Sect. 2.1) is used in our work to stay consistent with the original work by Vicient and Moreno [16].

Specifically, the similarity between two hashtags, h_i and h_j, is calculated as the maximum pairwise similarity (based on the Wu-Palmer measure) between one set of concepts in LC_{h_i} and another set of concepts in LC_{h_j}. Calculating the similarity this way is expected to find the correct sense of hashtag (among all the sense/concepts in LC_h).

Finally, in the third stage (i.e., clustering), any clustering algorithm can be used to cluster hashtags based on the similarity matrix obtained in the second stage. As mentioned earlier, in this paper we use hierarchical clustering which was used in the original work by Vicient and Moreno [16].

4 Sense-Level Semantic Hashtag Clustering

In this section, we describe the enhancement made to the word-level semantic hashtag clustering and showcase its positive impact using a toy example. Both Stage 1 (i.e., semantic grounding) and Stage 3 (i.e., clustering) of the sense-level semantic clustering algorithm are essentially the same as those in the word-level semantic clustering algorithm (see Algorithm 1 in Sect. 3). So, here, we discuss only Stage 2 (i.e., similarity matrix construction) of the algorithm, with a focus on the difference in the calculation of maximum pairwise similarity.

4.1 Similarity Matrix Construction

Word-Level Versus Sense-Level Similarity Matrix. As mentioned in Sect. 3, the similarity between two hashtags h_i and h_j is defined as the maximum pairwise similarity between one set of senses in LC_{h_i} and another set of senses in LC_{h_j}. (Recall that LC_h denotes a list of senses retrieved from Wordnet to semantically ground a hashtag h.) This maximum pairwise similarity is an effective choice for disambiguating the sense of a hashtag and was used to achieve a positive effect in the word-based approach by Vicient and Moreno [16].

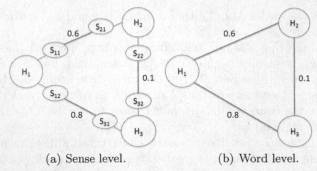

(a) Sense level.   (b) Word level.

(Edge weights denote similarity values (similarity = 1− distance). Assume the minimum similarity threshold is 0.5. Then, at the sense level (a), *two* clusters ($\{H_1, H_2\}, \{H_1, H_3\}$) should be formed because H_2 and H_3 are not similar (note $0.1 < 0.5$), but, at the word level (b), *one* cluster $\{H_1, H_2, H_3\}$ is formed because it appears as if H_2 and H_3 were similar via H_1. Moreover, the false triangle that appears to be formed at the word level violates the triangular inequality property because $\mathrm{dist}(H_1, H_2) + \mathrm{dist}(H_1, H_3) < \mathrm{dist}(H_2, H_3)$.)

Fig. 2. An illustration of clustering at the word level versus sense level.

However, we have observed many instances where a hashtag word is polysemic (i.e., has multiple senses) and it introduces an error in the clustering result. That is, the word-level algorithm does not distinguish among different senses of the same word when constructing a similarity matrix and, as a result, two hashtags are misjudged to be semantically similar (because they are similar to a third hashtag in two different senses) and are included in the same cluster. Moreover, a false triangle that violates the triangular inequality property may be formed at the word level. (Note this property is required of any distance metric like Wu-Palmer.) See Fig. 2 for an illustration. As its side effect, we have observed that a cluster tends to be formed centered around a hashtag that takes on multiple senses.

Thus, we chose to explicitly record the sense in which a hashtag is close to another hashtag when constructing a similarity matrix. This sense-level handling of hashtag semantic distance helps us ensure that the incorrect clustering problem of word-level clustering does not happen. Accordingly, it avoids the formation of clusters that are centered around a hashtag that has multiple senses.

Word-Level Similarity Matrix Construction. Algorithm 2 outlines the steps of calculating maximum pairwise similarity between hashtags in the word-level algorithm. One maximum pairwise similarity value is calculated for each pair of hashtags semantically grounded in the previous stage (i.e., Stage 1) and is entered into the similarity matrix. The similarity matrix size is $|H|^2$, where $|H|$ is the number of hashtags that have at least one sense (i.e., nonempty LC_h). Note that the pairwise similarity comparison is still done at the sense level, considering all senses of the hashtags that are compared.

Input: set H of hashtags h with nonempty LC_h.
Output: pairwise hashtag similarity matrix.

1 Initialize an empty similarity matrix $\mathbf{M}[|H|, |H|]$.
2 Initialize *maxSim* to 0.
3 **for** *each pair* (h_i, h_j) *of hashtags in* H **do**
4   // Calculate the maximum pairwise similarity between h_i and h_j.
5   **for** *each* $s_p \in LC_{h_i}$ **do**
6   **for** *each* $s_q \in LC_{h_j}$ **do**
7   Calculate the similarity *sim* between s_p and s_q.
8   **if** $sim > maxSim$ **then**
9 | Update *maxSim* to *sim* .
10   **end**
11   **end**
12   **end**
13   Enter *maxSim* into $\mathbf{M}[i, j]$.
14 **end**

Algorithm 2. Word-level construction of semantic similarity matrix.

Sense-Level Similarity Matrix Construction. Algorithm 3 outlines the steps of constructing a similarity matrix in the sense-level algorithm. Unlike the case of the word-level algorithm, entries in the similarity matrix are between senses that make maximum similarity pairs between a pair of hashtags. Since these senses are not known until the maximum pairwise similarity calculations are completed, the construction of the similarity matrix is deferred until then. In the first phase (Lines 2–16), for each pair of hashtags, the algorithm saves the pair of senses $(h_i.s_p,\ h_j.s_q)$ in the maximum similarity pair and the maximum similarity value in the list LH_s. Then, in the second phase (Lines 18–22), for each triplet element $(h_i.s_p,\ h_j.s_q,\ maxSim)$ in LH_s, the algorithm enters the maximum similarity value *maxSim* at the matrix index corresponding to the pair of senses $(h_i.s_p,\ h_j.s_q)$.

This two-phase construction of similarity matrix brings two advantages. First, it enables the algorithm to use exactly the needed number of matrix entries for those senses that are *distinct* among all senses that constitute pairwise maximum similarities between hashtags. The size of the matrix, therefore, is $|\hat{S}|^2$, where $\hat{S}$ is the set of distinct senses in LH_s (see Lines 18–19). Second, it enables the algorithm to add exactly the needed number of entries, that is, $|H|^2$ entries (i.e., one for each pair of hashtags (see Lines 20–22)) into a matrix of size $|\hat{S}|^2$, where $|\hat{S}|^2 > |H|^2$. (The remaining entries are initialized to 0 and remain 0, as they are for pairs of senses that do not represent maximum similarity pair between any hashtags.) Our observation is that the ratio $|\hat{S}|/|H|$ is limited to the range of 5 to 10 for most individual hashtags, which is consistent with Vicient's statement [15] that, out of semantically-grounded 903 hashtags, almost 100 of them have only 2 senses and very few have more than 5 senses.

Since what is clustered are *hashtags*, although their similarities are measured at the sense level, a number of interesting points hold. First, we do not need to add similarities between all pairs of senses in the similarity matrix. Second,

Input: set H of hashtags h with nonempty LC_h.
Output: pairwise hashtag similarity matrix.

1 Create an empty list LH_s of (hashtag sense pair, pairwise maximum similarity).
2 **for** *each pair (h_i, h_j) of hashtags in H* **do**
3   // `Calculate the maximum pairwise similarity between` h_i `and` h_j.
4   Initialize *maxSim* to 0.
5   Initialize *maxSimPair* to (null, null).
6   **for** *each $s_p \in LC_{h_i}$* **do**
7   **for** *each $s_q \in LC_{h_j}$* **do**
8   Calculate the similarity *sim* between s_p and s_q.
9   **if** *sim > maxSim* **then**
10   Update *maxSim* to *sim* .
11   Update *maxSimPair* to $(h_i.s_p, h_j.s_q)$.
12   **end**
13   **end**
14   **end**
15   Add *(maxSimPair, maxSim)* to LH_s.
16 **end**
17 // `Construct the similarity matrix.`
18 Count the number $|\hat{S}|$ of distinct hashtag senses in LH_s.
19 Initialize a similarity matrix $\mathbf{M}[|\hat{S}|, |\hat{S}|]$ as a **0** matrix.
20 **for** *each triplet $(h_i.s_p, h_j.s_q, maxSim)$ in LH_s* **do**
21   Update the $\mathbf{M}[m, n]$ to *maxSim*, where (m, n) is the matrix index for $(h_i.s_p, h_j.s_q)$.
22 **end**

Algorithm 3. Sense-level construction of semantic similarity matrix.

a hashtag may appear in multiple clusters, where each cluster is formed based on distinct senses of the hashtag, and therefore the resulting clusters are *overlapping*.

4.2 A Toy Example

To demonstrate the merit of clustering at the sense level as opposed to the word level, we made a toy set of hashtags and ran the metadata-based semantic clustering algorithm at both the word level and the sense level. The hashtags used are #date, #august, #tree, and #fruit. From Wordnet, we found that there were 3 senses associated with the word august, 13 senses with date, 5 senses with fruit, and 7 senses with tree.

Using the Wu-Palmer similarity measure (Sect. 2.1) at the word level, we obtained the distance matrix on the right. Then, to perform clustering using this distance matrix as the input, we used both the single-linkage and UPGMA (Sect. 2.3) as the distance

Hashtag	august	date	fruit	tree
august	0.000	0.200	0.500	0.667
date	0.200	0.000	0.100	0.400
fruit	0.500	0.100	0.00	0.556
tree	0.667	0.400	0.556	0.000

measure between newly formed clusters and set the distance threshold for extracting flat clusters from hierarchical clusters to 0.5.

Table 1 shows the clusters obtained using the word-level clustering. We see that #august, #date, and #fruit are included in the same cluster in both cases of the distance measure. This example demonstrates a case in which #date takes on multiple sense identities and glues together #august and #fruit in the same cluster at the word level although these two are not similar at the sense level, as shown next.

Table 1. Cluster assignment at word level.

Hashtag	Cluster using single-linkage	Cluster using UPGMA
august	1	1
date	1	1
fruit	1	1
tree	1	2

Now, using the sense-level clustering, out of a total of 28 senses associated with the four hashtags, the algorithm picked 10 senses shown in Table 2. These 10 senses were picked as a result of maximum pairwise similarity calculations between two sets of senses belonging to each pair of hashtags. (With 4 hashtags, there are a maximum of 12 ($= 2 \times 6$) senses that can be obtained for 6 ($= C(4, 2)$) maximum similarity pairs, and in this example case, there were duplicate senses, consequently giving 10 distinct senses.) As mentioned earlier, each of these senses represents the semantics of the hashtag word it belongs to, and thus makes an entry into the similarity (or distance) matrix input to the hierarchical clustering algorithm.

Table 2. Senses and their semantics (source: Wordnet).

Sense	Semantics
august.n.01	The month following July and preceeding September
august.a.01	Of or befitting a lord
corner.v.02	Force a person or animal into a position from which he can not escape
date.n.02	A participant in a date
date.n.06	The particular day, month, or year (usually according to Gregorian calendar) that an even occurred
date.n.08	Sweet edible fruit of the date palm with single long woody seed
fruit.n.01	The ripened reproductive body of a seed plant
fruit.v.01	Cause to bear fruit
tree.n.01	A tall perennial woody plant having a main trunk and branches forming a distinct elevated crown; includes both gymnosperms and angiosperms
yield.n.03	An amount of product

('n' stands for noun, 'v' for verb and 'a' for adjective.)

The distance matrix obtained from the 10 senses is shown in Fig. 3. The numbers in bold face are the maximum similarity values entered. Note that distance 1.000 means similarity 0.000.

Hashtag sense	Hashtag	august.n.01 august	august.a.01 august	corner.v.02 tree	date.n.02 date	date.n.06 date	date.n.08 date	fruit.n.01 fruit	fruit.v.01 fruit	tree.n.01 tree	yield.n.03 fruit
august.n.01	august	0.000	1.000	1.000	1.000	0.200	1.000	1.000	1.000	1.000	1.000
august.a.01	august	1.000	0.000	0.667	1.000	1.000	1.000	1.000	0.500	1.000	1.000
corner.v.02	tree	1.000	0.667	0.000	1.000	1.000	1.000	1.000	1.000	1.000	1.000
date.n.02	date	1.000	1.000	1.000	0.000	1.000	1.000	1.000	1.000	0.400	1.000
date.n.06	date	0.200	1.000	1.000	1.000	0.000	1.000	1.000	1.000	1.000	1.000
date.n.08	date	1.000	1.000	1.000	1.000	1.000	0.000	0.100	1.000	1.000	1.000
fruit.n.01	fruit	1.000	1.000	1.000	1.000	1.000	0.100	0.000	1.000	1.000	1.000
fruit.v.01	fruit	1.000	0.500	1.000	1.000	1.000	1.000	1.000	0.000	1.000	1.000
tree.n.01	tree	1.000	1.000	1.000	0.400	1.000	1.000	1.000	1.000	0.000	0.556
yield.n.03	tree	1.000	1.000	1.000	1.000	1.000	1.000	1.000	1.000	0.556	0.000

Fig. 3. Distance matrix in the toy example.

Table 3 shows the resulting cluster assignments. (The outcome is the same for both distance measures, which we believe is coincidental.) We see that #august and #date are together in the same cluster and so are #date and #fruit but, unlike the word-level clustering result, the three of #august, #date, and #fruit are not altogether in the same cluster. This separation is because, at the sense level, #date can no longer take on multiple identities as it did at the word level.

Table 3. Cluster assignment at the sense level.

Hashtag	Hashtag sense	Cluster using single-linkage	Cluster using UPGMA
date	date.n.02	1	1
tree	tree.n.01	1	1
fruit	yield.n.03	2	2
fruit	fruit.v.01	3	3
august	august.a.01	3	3
tree	corner.v.02	4	4
fruit	fruit.n.01	5	5
date	date.n.08	5	5
august	august.n.01	6	6
date	date.n.06	6	6

5   Evaluation

The focus of evaluating the sense-level clustering algorithm is on the accuracy gained from the finer granularity of semantics compared with the word-level clustering algorithm. To highlight this focus, we conducted two experiments distinguished by the choice of control on the semantics of hashtags used – the semantics are not controlled in one experiment and are controlled in the other.

All algorithms were implemented in Python and the experiments were performed on a computer with OS X operating system, 2.6 GHz Intel Core i5 processor, and 8 GB 1600 MHz DDR3 memory.

5.1   Experiment Setup

Performance Metric. We use *f-score*, which is commonly used in conjunction with recall and precision to evaluate clusters in reference to ground truth clusters, as the accuracy metric. In our evaluation, the f-score is calculated for each pair of a cluster in the ground truth cluster set and a cluster in the evaluated algorithm's output cluster set. Then, the final f-score resulting from the comparison of the two cluster sets is obtained in two different ways, depending on the purpose of the evaluation. For the purpose of evaluating individual output clusters, the pairwise maximum (i.e., "best match") f-score, denoted as f^m-score, is used as the final score. Given a ground truth cluster G_i matched against an output cluster set $\mathbf{C}$, the f^m-score is obtained as

$$f^m\text{-score}(G_i, \mathbf{C}) = \max_{C_j \in \mathbf{C} \,\wedge\, f\text{-score}(G_i, C_j) > 0} f\text{-score}(G_i, C_j) \qquad (4)$$

where the pairwise matching is one-to-one between $\mathbf{G}$ and $\mathbf{C}$.

On the other hand, for comparing overall accuracy of the entire set of clusters, the weighted average of pairwise maximum f-scores, denoted as f^a-score, is used instead. Given a ground truth cluster set $\mathbf{G}$ and an output cluster set $\mathbf{C}$, the f^a-score is calculated as

$$f^a\text{-score}(\mathbf{G}, \mathbf{C}) = \frac{\sum_{G_i \in \mathbf{G}} (f^m\text{-score}(G_i, \mathbf{C}) \times |G_i|)}{\sum_{G_i \in \mathbf{G}} |G_i|} \qquad (5)$$

Distance Threshold. The distance threshold for determining flat clusters in hierarchical clustering was set using the "best result" approach. That is, we tried both distance measures (i.e., single-linkage and UPGMA) and different distance threshold values and picked the measure and value that produced the best result based on the weighted average f-score measure.

Qualified Output Clusters. The clustering output shows a large number of small clusters, many of them including only one or two hashtags. Thus, for a given ground truth cluster, we consider the best matching output cluster only if it contains at least 3 hashtags and the f^m-score is greater than 0.1.

5.2   Experiment 1: Uncontrolled Hashtag Semantics

In this experiment, hashtags are collected without any control over the semantics, i.e., "indiscriminately", from tweet messages.

To build the ground truth, we manually gathered 2,910 tweets from the Symplur web site (www.symplur.com) – the same number of Symplur tweets was also used in the evaluation of word-level clustering by Vicient and Moreno [16]. There were 1,010 unique hashtags in the 2,910 tweets. We then manually annotated the semantics of the 1,010 hashtags to choose 230 hashtags and classified them into 15 clusters. The remaining hashtags were classified as noise. Figure 4 shows the sizes of the resulting ground truth clusters.

In the hierarchical clustering, distance threshold value that gave the best result was 0.4 when the UPGMA measure was used for both sense-level and word-level.

Figure 5 shows the accuracies achieved by semantic clustering at the word-level and the sense-level using the uncontrolled hashtag from the Symplur dataset. Table 4 shows more details, including precision and recall for individual

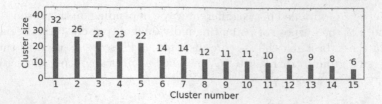

Fig. 4. Sizes of ground truth clusters in the uncontrolled hashtag experiment.

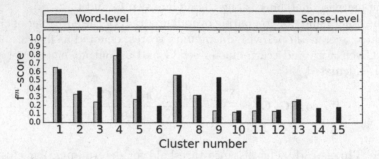

Fig. 5. Maximum pairwise f-scores of output clusters for word-level and sense-level in the uncontrolled hashtag experiment.

clusters. From the results we see that every sense-level cluster outperforms the word-level counterpart (except cluster 1 due to rounding-off difference). Particularly, the f^m-scores are zero for word-level clusters 6, 14, and 15, thus bringing the performance gain to "infinity". (Word-level clustering did not generate any qualified matching cluster for the ground truth clusters 6, 14, and 15.) Further, when all 15 clusters are considered together, the weighted average of maximum pairwise f-scores, f^a-score, is 0.43 for sense-level clustering and 0.34 for word-level clustering – a 26% gain.

5.3   Experiment 2: Controlled Hashtag Semantics

In this experiment, hashtags are collected with controlled semantics, that is, "discriminated" based on their semantics.

The ground truth clusters were prepared using a combination of two tweet datasets. One dataset has 2.5 million tweets collected through Twitter REST API. This dataset contained 708 hashtags that had 20 or more tweets associated with them. We selected from the 708 hashtags approximately 150 hashtags based on the topics they addressed. Then, we started with the 15 clusters in the ground truth from the Symplur dataset used in Experiment 1 (see Sect. 5.2) and randomly picked from them one cluster at a time, merging them, until the total number of hashtags in the selected clusters reached approximately 150. Then, we combined the two sets of approximately 150 hashtags each to form one set, which contained a total of 309 hashtags. Figure 6 shows the sizes of the resulting ground truth clusters.

Table 4. Details of gold standard test results in the uncontrolled hashtag experiment.

Ground truth clusters		Sense-level clusters				Word-level clusters			
Id	Size	Recall	Precision	f^m-score	Size	Recall	Precision	f^m-score	Size
1	32	0.63	0.65	0.63	31	0.63	0.67	0.65	30
2	26	0.35	0.39	0.37	23	0.31	0.35	0.33	23
3	23	0.39	0.43	0.41	21	0.35	0.19	0.24	43
4	23	0.91	0.84	0.88	25	0.83	0.76	0.79	25
5	22	0.41	0.45	0.43	20	0.41	0.20	0.27	44
6	14	0.21	0.18	0.19	17	n/a	n/a	n/a	n/a
7	14	0.64	0.50	0.56	18	0.64	0.50	0.56	18
8	12	0.25	0.43	0.32	7	0.50	0.24	0.32	25
9	11	0.82	0.39	0.53	23	0.82	0.08	0.14	118
10	11	0.18	0.11	0.14	18	0.09	0.17	0.12	6
11	10	0.40	0.27	0.32	15	0.50	0.08	0.14	59
12	9	0.11	0.25	0.15	4	0.11	0.17	0.13	6
13	9	0.22	0.33	0.27	6	0.22	0.29	0.25	7
14	8	0.13	0.25	0.17	4	n/a	n/a	n/a	n/a
15	6	0.17	0.20	0.18	5	n/a	n/a	n/a	n/a

f^a-score is 0.43 for sense-level clusters and 0.34 for word-level clusters.

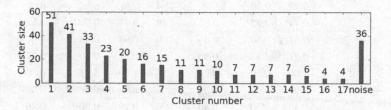

Fig. 6. Sizes of ground truth clusters in the controlled hashtag experiment.

For the hierarchical clustering, the distance threshold values that gave the best results were 0.3 at the word level and 0.4 at the sense level, respectively, both using the UPGMA measure.

Figure 7 shows the accuracies achieved by the semantic clustering at the word-level and the sense-level using the controlled hashtags. Table 5 shows more details. Sense-level clusters outperform word-level clusters in 12 of 17 clusters and are very close runner-up's in the remaining 5 clusters. Compared with Experiment 1, the 309 hashtags with controlled semantics gave sense-level clustering 1,293 unique senses to work with to produce best match clusters with higher f^m-scores to a larger number of ground truth clusters. In contrast, word-level clustering did not generate any qualified matching cluster for the ground truth clusters 6, 11, 15, 16 and 17, and hence the accuracy suffered significantly. Consequently, when all 17 clusters are considered, the f^a-score is 0.50 for sense-level clustering and 0.34 for word-level – a gain of 47%. Notably, in this Experiment 2,

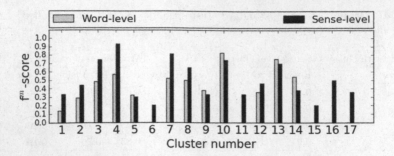

Fig. 7. Maximum pairwise f-scores of output clusters for word-level and sense-level in the controlled hashtag experiment.

Table 5. Details of gold standard test results in the controlled hashtag experiment.

Ground truth clusters		Sense-level clusters				Word-level clusters			
Id	Size	Recall	Precision	f^m-score	Size	Recall	Precision	f^m-score	Size
1	51	0.25	0.45	0.33	29	0.08	0.80	0.14	5
2	41	0.29	0.86	0.44	14	0.20	0.57	0.29	14
3	33	0.61	0.95	0.74	21	0.39	0.62	0.48	21
4	23	0.91	0.95	0.93	22	0.52	0.63	0.57	19
5	20	0.30	0.30	0.30	20	0.20	0.80	0.32	5
6	16	0.13	0.50	0.20	4	n/a	n/a	n/a	n/a
7	15	0.73	0.92	0.81	12	0.47	0.58	0.52	12
8	11	0.91	0.50	0.65	20	0.64	0.41	0.50	17
9	11	0.27	0.43	0.33	7	0.27	0.60	0.37	5
10	10	1.00	0.59	0.74	17	1.00	0.71	0.83	14
11	7	0.29	0.40	0.33	5	n/a	n/a	n/a	n/a
12	7	0.43	0.50	0.46	6	0.29	0.50	0.36	4
13	7	1.00	0.54	0.70	13	0.86	0.67	0.75	9
14	7	0.43	0.33	0.38	9	0.43	0.75	0.55	4
15	6	0.17	0.25	0.20	4	n/a	n/a	n/a	n/a
16	4	0.50	0.50	0.50	4	n/a	n/a	n/a	n/a
17	4	0.50	0.29	0.36	7	n/a	n/a	n/a	n/a

f^a-score is 0.50 for sense-level clusters and 0.34 for word-level clusters.

the sense-level outperforms the word-level by a substantially larger margin than in Experiment 1 because hashtags were hand-picked deliberately based on their semantics, while the word-level performs the same as in Experiment 1 because word-level is "oblivious" to the exact (i.e., sense-level) semantics of hashtags.

6 Related Work

There are several works on semantic clustering of hashtags that focused on the contextual semantics of hashtags [6,8,10,12,13] by using the bag of words model to represent the texts accompanying a hasthag. Tsur et al. [12,13] and Muntean et al. [6] appended tweets that belonged to each unique hashtag into a unique document called "virtual document". These documents were then represented as vectors in the vector space model. Rosa et al. [8] used hashtag clusters to achieve topical clustering of tweets, where they compared the effects of expanding URLs found in tweets. Stilo and Paola [10] clustered hashtag "senses" based on their temporal co-occurrence with other hashtags. The term "sense" in their work is different from the lexical sense used in this paper.

Lacking the ability to form lexical semantic sense-level clusters of hashtag has been a major shortcoming of the current approaches. To the best our knowledge, the work by Vicient and Moreno [16] is the only one that opened research in this direction. They used Wordnet and Wikipedia as the metadata source for clustering hashtags at the word level.

7 Conclusion

In this paper, we enhanced the current metadata-based semantic hashtag clustering algorithm by determining semantic similarity between hashtags at the *sense* level as opposed to the word level. This sense-level decision on clustering avoids incorrectly putting hashtags of different senses in the same cluster. The result was significantly higher accuracy of semantic clusters without increasing the complexities of the algorithm in practice. Gold standard tests showed an overall gain of 26% (in case of uncontrolled hashtag semantics) and 47% (in case of controlled hashtag semantics) in the weighted average of maximum pairwise f-scores.

For the future work, new metadata sources can be added to provide the metadata-based semantic hashtag clustering algorithm with more abilities. For example, a crowdsourced website like Urban Dictionary (www.urbandictionary.com) that specializes in informal human communication can be a helpful metadata source for decoding lexical semantics of hashtags. Internet search engines also provide rich information on the semantics of hashtags. Additionally, online translation service like Google Translate (translate.google.com) can be a good source for understanding hashtags of a different language as well as correcting spelling errors and expanding abbreviations.

References

1. Bhulai, S., et al.: Trend visualization on twitter: what's hot and what's not? In: 1st International Conference on Data Analytics, pp. 43–48 (2012)
2. Costa, J., Silva, C., Antunes, M., Ribeiro, B.: Defining semantic meta-hashtags for Twitter classification. In: Tomassini, M., Antonioni, A., Daolio, F., Buesser, P. (eds.) ICANNGA 2013. LNCS, vol. 7824, pp. 226–235. Springer, Heidelberg (2013). doi:10.1007/978-3-642-37213-1_24
3. Javed, A., Lee, B.S.: Sense-level semantic clustering of hashtags in social media. In: the 3rd Annual International Symposium on Information Management and Big Data, September 2016

4. Kelly, R.: Twitter study reveals interesting results about usage - 40% is point-less babble. http://pearanalytics.com/blog/2009/twitter-study-reveals-interesting-results-40-percent-pointless-babble/, Accessed 05 Oct 2016
5. Manning, C.D., Raghavan, P., Schütze, H.: Introduction to Information Retrieval. Cambridge University Press (2008). Chap. 17
6. Muntean, C.I., Morar, G.A., Moldovan, D.: Exploring the meaning behind Twitter hashtags through clustering. In: Abramowicz, W., Domingue, J., Węcel, K. (eds.) BIS 2012. LNBIP, pp. 231–242. Springer, Heidelberg (2012). doi:10.1007/978-3-642-34228-8_22
7. Park, S., Shin, H.: Identification of implicit topics in twitter data not containing explicit search queries. In: 25th International Conference on Computational Linguistics, pp. 58–68 (2014)
8. Rosa, K.D., Shah, R., Lin, B.: Topical clustering of tweets. In: 3rd Workshop on Social Web Search and Mining. pp. 133–138, July 2011
9. Saif, H., He, Y., Alani, H.: Semantic sentiment analysis of Twitter. In: Cudré-Mauroux, P., Heflin, J., Sirin, E., Tudorache, T., Euzenat, J., Hauswirth, M., Parreira, J.X., Hendler, J., Schreiber, G., Bernstein, A., Blomqvist, E. (eds.) ISWC 2012. LNCS, vol. 7649, pp. 508–524. Springer, Heidelberg (2012). doi:10.1007/978-3-642-35176-1_32
10. Stilo, G., Velardi, P.: Temporal semantics: time-varying hashtag sense clustering. In: 19th International Conference on Knowledge Engineering and Knowledge Management, pp. 563–578, November 2014
11. Teu, P., Kraxberger, S.: Extracting semantic knowledge from Twitter. In: 3rd IFIP WG 8.5 International Conference on Electronic Participation, pp. 48–59 (2011)
12. Tsur, O., Littman, A., Rappoport, A.: Efficient clustering of short messages into general domains. In: 7th International AAAI Conference on Weblogs and Social Media (2013)
13. Tsur, O., Littman, A., Rappoport, A.: Scalable multi stage clustering of tagged micro-messages. In: International Conference on World Wide Web. pp. 621–622, April 2012
14. Usage Statistics. http://www.internetlivestats.com/twitter-statistics/
15. Vicient, C.: Moving towards the semantic web: enabling new technologies through the semantic annotation of social contents. PhD thesis. Universitat Robira I Virgili, December 2014
16. Vicient, C., Moreno, A.: Unsupervised semantic clustering of twitter hashtags. In: 21st European Conference on Artificial Intelligence, pp. 1119–1120, August 2014
17. Wang, X., et al.: Topic sentiment analysis in twitter: a graph-based hashtag sentiment classification approach. In: 20th ACM Conference on Information and Knowledge Management, pp. 1031–1040 (2011)
18. Zhibiao, W., Palmer, M.: Verbs semantics and lexical selection. In: 32nd Annual Meeting on Association for Computational Linguistics, pp. 133–138 (1994)

Automatic Idiom Recognition with Word Embeddings

Jing Peng$^{(\boxtimes)}$ and Anna Feldman

Department of Computer Science and Department of Linguistics,
Montclair State University, Montclair, NJ 07043, USA
pengj@mail.montclair.edu

Abstract. Expressions, such as *add fuel to the fire*, can be interpreted literally or idiomatically depending on the context they occur in. Many Natural Language Processing applications could improve their performance if idiom recognition were improved. Our approach is based on the idea that idioms and their literal counterparts do not appear in the same contexts. We propose two approaches: (1) Compute inner product of context word vectors with the vector representing a target expression. Since literal vectors predict well local contexts, their inner product with contexts should be larger than idiomatic ones, thereby telling apart literals from idioms; and (2) Compute literal and idiomatic scatter (covariance) matrices from local contexts in word vector space. Since the scatter matrices represent context distributions, we can then measure the difference between the distributions using the Frobenius norm. For comparison, we implement [8,16,24] and apply them to our data. We provide experimental results validating the proposed techniques.

Keywords: Idiom recognition · Vector space models · Distributional semantics · Word embeddings

1 Introduction

Natural language is filled with emotion and implied intent, which are often not trivial to detect. One specific challenge are idioms. Figurative language draws off of prior references and is unique to each culture and sometimes what we don't say is even more important than what we do. This, naturally, presents a significant problem for many Natural Language Processing (NLP) applications as well as for big data analytics.

Idioms are conventionalized expressions whose figurative meanings cannot be derived from literal meaning of the phrase. There is no single agreed-upon definition of idioms that covers all members of this class [3,10,13,19,22,25]. At the same time, idioms do not form a homogeneous class that can be easily defined. Some examples of idioms are *I'll eat my hat* (I'm confident), *Cut it out* (Stop talking/doing something), *a blessing in disguise* (some bad luck or misfortune results in something positive), *kick the bucket* (die), *ring a bell* (sound

© Springer International Publishing AG 2017
J.A. Lossio-Ventura and H. Alatrista-Salas (Eds.): SIMBig 2015/2016, CCIS 656, pp. 17–29, 2017.
DOI: 10.1007/978-3-319-55209-5_2

familiar), *keep your chin up* (remain cheerful), *piece of cake* (easy task), *miss the boat* (miss out on something), *(to be) on the ball* (be attentive/competent), *put one's foot in one's mouth* (say something one regrets), *rake someone over the coals* (to reprimand someone severely), *under the weather* (sick), *a hot potato* (controversial issue), *an arm and a leg* (expensive), *at the drop of a hat* (without any hesitation), *barking up the wrong tree* (looking in the wrong place), *beat around the bush* (avoiding main topic).

It turns out that expressions are often ambiguous between an idiomatic and a literal interpretation, as one can see in the examples below[1]:

(A) After the last page was sent to the printer, an editor would **ring a bell**, walk toward the door, and holler "Good night!" (Literal) (B) His name never fails to **ring a bell** among local voters. Nearly 40 years ago, Carthan was elected mayor of Tchula... (Idiomatic)

(C) ... that caused the reactor to literally **blow its top**. About 50 tons of nuclear fuel evaporated in the explosion... (Literal) (D) ... He didn't pound the table, he didn't **blow his top**. He always kept his composure. (Idiomatic)

(E) ... coming out of the fourth turn, slid down the track, **hit** the inside **wall** and then hit the attenuator at the start of pit road. (Literal) (F) ... job training, research and more have **hit** a Republican **wall**. (Idiomatic)

[8] analysis of 60 idioms from the British National Corpus (BNC) has shown that close to half of these also have a clear literal meaning; and of those with a literal meaning, on average around 40% of their usages are literal. Therefore, idioms present great challenges for many Natural Language Processing (NLP) applications. Most current translation systems rely on large repositories of idioms. Unfortunately, more frequently than not, MT systems are not able to translate idiomatic expressions correctly.

In this paper we describe an algorithm for automatic classification of idiomatic and literal expressions. Similarly to [21], we treat idioms as semantic outliers. Our assumption is that the context word distribution for a literal expression will be different from the distribution for an idiomatic one. We capture the distribution in terms of covariance matrix in vector space.

2   Our Approach

Our idea is simple: idiomatic expressions and their literal counterparts do not occur in the same contexts. We formulate two hypotheses.

1. Projection Based on Local Context Representation
   We hypothesize that words in a given text segment that are representatives of the local context are likely to associate strongly with a literal expression in the segment, in terms of projection (or inner product) of word vectors onto the vector representing the literal expression.

---

[1] These examples are extracted from the Corpus of Contemporary American English (COCA) (http://corpus.byu.edu/coca/).

2. Local Context Distributions

We hypothesize that the context word distribution for a literal expression in word vector space will be different from the distribution for an idiomatic one. This hypothesis also underlies the distributional approach to meaning [11,15].

We want to emphasize that our approach is applicable to any type of syntactic constructions, but the experiments described below are based on the data released by [8], i.e., verb-noun constructions (VNCs). Thus, we can directly compare the performance of our model to [8] work.

2.1 Projection Based on Local Context Representation

The local context of a literal target verb-noun construction (VNC) must be different from that of an idiomatic one. We propose to exploit recent advances in vector space representation to capture the difference between local contexts [17,18].

A word can be represented by a vector of fixed dimensionality q that best predicts its surrounding words in a sentence or a document [17,18]. Given such a vector representation, our first proposal is the following. Let v and n be the vectors corresponding to the verb and noun in a target verb-noun construction, as in *blow whistle*, where $v \in \Re^q$ represents *blow* and $n \in \Re^q$ represents *whistle*. Let $\sigma_{vn} = v + n \in \Re^q$. Thus, σ_{vn} is the word vector that represents the composition of verb v and noun n, and in our example, the composition of *blow* and *whistle*. As indicated in [18], word vectors obtained from deep learning neural net models exhibit linguistic regularities, such as additive compositionality. Therefore, σ_{vn} is justified to predict surrounding words of the composition of, say, *blow* and *whistle*. Our hypothesis is that on average, inner product $\sigma_{blowwhistle} \cdot v$, where vs are context words in a literal usage, should be greater than $\sigma_{blowwhistle} \cdot v$, where vs are context words in an idiomatic usage.

For a given vocabulary of m words, represented by matrix

$$V = [v_1, v_2, \cdots, v_m] \in \Re^{q \times m}, \tag{1}$$

we calculate the projection of each word v_i in the vocabulary onto σ_{vn}

$$P = V^t \sigma_{vn} \tag{2}$$

where $P \in \Re^m$, and t represents transpose. Here we assume that σ_{vn} is normalized to have unit length. Thus, $P_i = v_i^t \sigma_{vn}$ indicates how strongly word vector v_i is associated with σ_{vn}. This projection, or inner product, forms the basis for our proposed technique.

Let

$$D = \{d_1, d_2, \cdots, d_l\}$$

be a set of l text segments (local contexts), each containing a target VNC (i.e., σ_{vn}). Instead of generating a term by document matrix, where each term is *tf-idf* (product of term frequency and inverse document frequency), we compute a term by document matrix $M_D \in \Re^{m \times l}$, where each term in the matrix is

$$p \cdot idf, \tag{3}$$

the product of the projection of a word onto a target VNC and inverse document frequency. That is, the term frequency (tf) of a word is replaced by the projection (inner product) of the word onto σ_{vn} (2). Note that if segment d_j does not contain word v_i, $M_D(i,j) = 0$, which is similar to *tf-idf* estimation. The motivation is that topical words are more likely to be well predicted by a literal VNC than by an idiomatic one. The assumption is that a word vector is learned in such a way that it best predicts its surrounding words in a sentence or a document [17,18]. As a result, the words associated with a literal target will have larger projection onto a target σ_{vn}. On the other hand, the projections of words associated with an idiomatic target VNC onto σ_{vn} should have a smaller value.

We also propose a variant of $p \cdot idf$ representation. In this representation, each term is a product of p and typical *tf-idf*. That is,

$$p \cdot tf \cdot idf. \tag{4}$$

2.2 Local Context Distributions

Our second hypothesis states that words in a local context of a literal expression will have a different distribution from those in the context of an idiomatic one. We propose to capture local context distributions in terms of scatter matrices in a space spanned by word vectors [17,18].

Let $d = (w_1, w_2 \cdots, w_k) \in \Re^{q \times k}$ be a segment (document) of k words, where $w_i \in \Re^q$ are represented by a vectors [17,18]. Assuming w_is have been centered, we compute the scatter matrix

$$\Sigma = d^t d, \tag{5}$$

where Σ represents the local context distribution for a given target VNC.

Given two distributions represented by two scatter matrices Σ_1 and Σ_2, a number of measures can be used to compute the distance between Σ_1 and Σ_2, such as Choernoff and Bhattacharyya distances [12]. Both measures require the knowledge of matrix determinant. In our case, this can be problematic, because Σ (5) is most likely to be singular, which would result in a determinant to be zero.

We propose to measure the difference between Σ_1 and Σ_2 using matrix norms. We have experimented with the Frobenius norm and the spectral norm. The Frobenius norm evaluates the difference between Σ_1 and Σ_2 when they act on a standard basis. The spectral norm, on the other hand, evaluates the difference when they act on the direction of maximal variance over the whole space.

3 Data Preprocessing

We use BNC [2] and a list of verb-noun constructions (VNCs) extracted from BNC by [6,8] and labeled as L (Literal), I (Idioms), or Q (Unknown).

The list contains only those VNCs whose frequency was greater than 20 and that occurred at least in one of two idiom dictionaries [7,23]. The dataset consists of 2,984 VNC tokens. For our experiments we only use VNCs that are annotated as I or L. We only experimented with idioms that can have both literal and idiomatic interpretations. We should mention that our approach can be applied to any syntactic construction. We decided to use VNCs only because this dataset was available and for fair comparison – most work on idiom recognition relies on this dataset.

We use the original SGML annotation to extract paragraphs from BNC. Each document contains three paragraphs: a paragraph with a target VNC, the preceding paragraph and following one.

Since BNC did not contain enough examples, we extracted additional ones from COCA, COHA and GloWbE (http://corpus.byu.edu/). Two human annotators labeled this new dataset for idioms and literals. The inter-annotator agreement was relatively low (Cohen's kappa = .58); therefore, we merged the results keeping only those entries on which the two annotators agreed.

Table 1. Datasets: Is = idioms; Ls = literals

Expression	Train	Test
BlowWhistle	20 Is, 20 Ls	7 Is, 31 Ls
LoseHead	15 Is, 15 Ls	6 Is, 4 Ls
MakeScene	15 Is, 15 Ls	15 Is, 5 Ls
TakeHeart	15 Is, 15 Ls	46 Is, 5 Ls
BlowTop	20 Is, 20 Ls	8 Is, 13 Ls
BlowTrumpet	50 Is, 50 Ls	61 Is, 186 Ls
GiveSack	20 Is, 20 Ls	26 Is, 36 Ls
HaveWord	30 Is, 30 Ls	37 Is, 40 Ls
HitRoof	50 Is, 50 Ls	42 is, 68 Ls
HitWall	90 Is, 90 Ls	87 is, 154 Ls
HoldFire	20 Is, 20 Ls	98 Is, 6 Ls
HoldHorse	80 Is, 80 Ls	162 Is, 79 Ls

4   Experiments

We have carried out an empirical study evaluating the performance of the proposed techniques. The goal is to predict the idiomatic usage of VNCs.

4.1   Methods

For comparison, the following methods are evaluated.

1. $tf \cdot idf$: compute term by document matrix from training data with $tf \cdot idf$ weighting.
2. $p \cdot idf$: compute term by document matrix from training data with proposed $p \cdot idf$ weighting (3).
3. $p \cdot tf \cdot idf$: compute term by document matrix from training data with proposed p * tf-idf weighting (4).
4. $CoVAR_{Fro}$: proposed technique (5) described in Sect. 2.2, the distance between two matrices is computed using Frobenius norm.
5. $CoVAR_{Sp}$: proposed technique similar to $CoVAR_{Fro}$. However, the distance between two matrices is determined using the spectral norm.
6. $Context+$ $(CTX+)$: supervised version of the CONTEXT technique described in [8] (see below).

For methods from **1** to **3**, we compute a latent space from a term by document matrix obtained from the training data that captures 80% variance. To classify a test example, we compute cosine similarity between the test example and the training data in the latent space to make a decision.

For methods **4** and **5**, we compute literal and idiomatic scatter matrices from training data (5). For a test example, compute a scatter matrix according to (5), and calculate the distance between the test scatter matrix and training scatter matrices using the Frobenius norm for method **4**, and the spectral norm for method **5**.

Method **6** corresponds to a supervised version of CONTEXT described in [8]. CONTEXT is unsupervised because it does not rely on manually annotated training data, rather it uses knowledge about automatically acquired canonical forms (C-forms). C-forms are fixed forms corresponding to the syntactic patterns in which the idiom normally occurs. Thus, the gold-standard is "noisy" in CONTEXT. Here we provide manually annotated training data. That is, the gold-standard is "clean." Therefore, CONTEXT+ is a supervised version of CONTEXT. We implemented this approach from scratch since we had no access to the code and the tools used in the original article and applied this method to our dataset and the performance results are reported in Table 3.

4.2   Word Vectors

For our experiments reported here, we obtained word vectors using the word2vec tool [17,18] and the text8 corpus. The text8 corpus has more than 17 million words, which can be obtained from mattmahoney.net/dc/text8.zip. The following shows a sample text8 corpus:

> anarchism originated as a term of abuse first used against early working class radicals including the diggers of the english revolution and the sans culottes of the french revolution whilst the term is still used in a pejorative way to describe any act that used violent means to destroy the organization of society it has also been taken up as a positive label by self defined anarchists the word anarchism is derived from the greek

without archons ruler chief king anarchism as a political philosophy is the belief that rulers are unnecessary and should be abolished although there are differing interpretations of what this means anarchism also refers to related social movements that advocate the elimination of authoritarian institutions particularly the state the word anarchy as most anarchists use it does not imply chaos nihilism or anomie but rather a harmonious anti authoritarian society in place of what are regarded as authoritarian political structures and coercive economic institutions anarchists advocate social relations based upon voluntary association of autonomous individuals mutual aid

The resulting vocabulary has 71,290 words, each of which is represented by a $q = 200$ dimension vector. Thus, this 200 dimensional vector space provides a basis for our experiments.

4.3 Datasets

Table 1 describes the datasets we used to evaluate the performance of the proposed technique. All these verb-noun constructions are ambiguous between literal and idiomatic interpretations. The examples below (from the corpora we used) show how these expressions can be used *literally*.

BlowWhistle: *we can immediately turn towards a high-pitched sound such as whistle being blown. The ability to accurately locate a noise is particularly important for the animals which use sound to find their way around.*

LoseHead: *Here are several degrees of deception. The simplest involves displaying a large eye-spot marking somewhere at the rear end of the body. This looks as eye-like to the predator as the real eye and gives the prey a fifty-fifty chance of losing its head. That was a very nice bull 1 shot, but I lost his head.*

MakeScene: *In another analogy our mind can be thought of as a huge tapestry in which the many episodes of life were originally isolated and there was no relationship between the parts, but at last we must make a unified scene of our whole life.*

TakeHeart: *He sacrifices four lambs ... then takes two inside and kills them by slitting open the throat and the chest and cutting off one of the forelegs at the shoulder so the heart can be taken out still pumping and offered to the god on a plate.*

BlowTop: *Yellowstone has no large sources of water to create the amount of steam to blow its top as in previous eruptions.*

5 Results

Table 2 shows the average precision, recall and accuracy of the competing methods on **BlowWhistle, LoseHead, MakeScene, BlowTop, BlowTrumpet**, and **GiveSack** over 20 runs. Table 3 shows the performance on **HitRoof, HitWall, HoldFire, TakeHeart, HaveWord**, and **HoldHorse**. The best performance is in bold face. The best model is identified by considering precision,

Table 2. Average precision, recall, and accuracy by each method on **BlowWhistle**, **LoseHead**, **MakeScene**, **BlowTop**, **BlowTrumpet**, and **GiveSack** datasets.

Method	BlowWhistle			LoseHead			MakeScene		
	Pre	Rec	Acc	Pre	Rec	Acc	Pre	Rec	Acc
$tf \cdot idf$	0.23	0.75	0.42	0.27	0.21	0.49	0.41	0.13	0.33
$p \cdot idf$	0.29	0.82	0.60	0.49	0.27	0.48	0.82	0.48	0.53
$p \cdot tf \cdot idf$	0.23	0.99	0.37	0.31	0.30	0.49	0.40	0.11	0.33
$CoVAR_{Fro}$	**0.65**	**0.71**	**0.87**	0.60	0.78	0.58	**0.84**	**0.83**	**0.75**
$CoVAR_{sp}$	0.44	0.77	0.77	**0.62**	**0.81**	**0.61**	0.80	0.82	0.72
$CTX+$	0.17	0.56	0.40	0.55	0.52	0.46	0.78	0.0.37	0.45
	BlowTop			BlowTrumpet			GiveSack		
	Pre	Rec	Acc	Pre	Rec	Acc	Pre	Rec	Acc
$tf \cdot idf$	0.55	0.93	0.65	0.26	0.85	0.36	0.61	0.63	0.55
$p \cdot idf$	0.59	0.58	0.68	0.44	0.85	0.69	0.55	0.47	0.62
$p \cdot tf \cdot idf$	0.54	0.53	0.65	0.33	0.93	0.51	0.54	0.64	0.55
$CoVAR_{Fro}$	**0.81**	**0.87**	**0.86**	0.45	0.94	0.70	0.63	0.88	0.72
$CoVAR_{sp}$	0.71	0.79	0.79	0.39	0.89	0.62	0.66	0.75	0.73
$CTX+$	0.66	0.70	0.75	**0.59**	**0.81**	**0.81**	**0.67**	**0.83**	**0.76**

Table 3. Average precision, recall, and accuracy by each method on **HitRoof**, **HitWall**, **HoldFire**, **TakeHeart**, **HaveWord**, and **HoldHorse** datasets.

Method	HitRoof			HitWall			HoldFire		
	Pre	Rec	Acc	Pre	Rec	Acc	Pre	Rec	Acc
$tf \cdot idf$	0.42	0.70	0.52	0.37	0.99	0.39	0.91	0.57	0.57
$p \cdot idf$	0.54	0.84	0.66	0.55	0.92	0.70	0.97	0.83	0.81
$p \cdot tf \cdot idf$	0.41	0.98	0.45	0.39	0.97	0.43	**0.95**	**0.89**	**0.85**
$CoVAR_{Fro}$	**0.61**	**0.88**	**0.74**	**0.59**	**0.94**	**0.74**	0.97	0.86	0.84
$CoVAR_{sp}$	0.54	0.85	0.66	0.50	0.95	0.64	0.96	0.87	0.84
$CTX+$	0.55	0.82	0.67	0.92	0.57	0.71	0.97	0.64	0.64
	TakeHeart			HaveWord			HoldHorse		
	Pre	Rec	Acc	Pre	Rec	Acc	Pre	Rec	Acc
$tf \cdot idf$	0.65	0.02	0.11	0.52	0.33	0.52	0.79	0.98	0.80
$p \cdot idf$	0.90	0.43	0.44	0.52	0.53	0.54	0.86	0.81	0.78
$p \cdot tf \cdot idf$	0.78	0.11	0.18	0.53	0.53	0.53	0.84	0.97	0.86
$CoVAR_{Fro}$	0.95	0.61	0.62	0.58	0.49	0.58	**0.86**	**0.97**	**0.87**
$CoVAR_{sp}$	0.94	0.55	0.56	0.56	0.53	0.58	0.77	0.85	0.73
$CTX+$	**0.92**	**0.66**	**0.64**	**0.53**	**0.85**	**0.57**	0.93	0.89	0.88

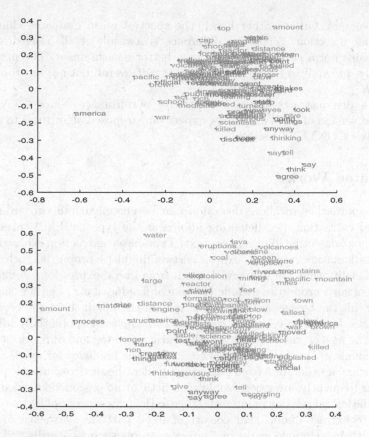

Fig. 1. Top: Projection of context words onto the subspace spanned by the first two eigenvectors of the idiomatic scatter. Bottom: Projection of context words onto the subspace of the literal scatter matrix. The blue words are idiomatic context, while the red words are literal context. (Color figure online)

recall, and accuracy together for each model. We calculate accuracy by adding true positives (idioms) and true negatives (literals) and normalizing the sum by the number of examples.

Figure 1 shows the projection of context words of **blow top** onto a subspace spanned by the first two eigenvectors of the scatter matrices (5). The top plot in Fig. 1 shows the projection onto the subspace of the idiomatic scatter matrix, while the lower plot shows the projection of the literal scatter matrix. The blue indicates idiomatic context words, while the red indicates literal context words. This shows that the scatter matrices (5) seem capable of capturing the difference in distributions between idiomatic context words and literal ones.

Interestingly, the Frobenius norm outperforms the spectral norm. One possible explanation is that the spectral norm evaluates the difference when two matrices act on the maximal variance direction, while the Frobenius norm evaluates on a standard basis. That is, Frobenius measures the difference along all basis vectors. On the other hand, the spectral norm evaluates changes in a particular direction. When the difference is a result of all basis directions, the Frobenius norm potentially provides a better measurement. The projection methods ($p \cdot idf$ and $p \cdot tf \cdot idf$) outperform $tf \cdot idf$ overall but not as pronounced as $CoVAR$.

$CTX+$ demonstrates a very competitive performance. Since $CTX+$ is a supervised version of CONTEXT, we expect our proposed algorithms to outperform Fazly's CONTEXT method.

6   Related Work

Previous approaches to idiom detection can be classified into two groups: (1) type-based extraction, i.e., detecting idioms at the type level; (2) token-based detection, i.e., detecting idioms in context. Type-based extraction is based on the idea that idiomatic expressions exhibit certain linguistic properties such as non-compositionality that can distinguish them from literal expressions [8,22]. Some examples of such properties include (1) lexical fixedness: e.g., neither 'shoot the wind' nor 'hit the breeze' are valid variations of the idiom shoot the breeze and (2) syntactic fixedness: e.g., *The guy kicked the bucket* is potentially idiomatic whereas *The bucket was kicked* is not idiomatic anymore; and of course, (3) non-compositionality. Thus, some approaches look at the tendency for words to occur in one particular order, or a fixed pattern. [14] identifies lexico-syntactic patterns that occur frequently, are recognizable with little or no precoded knowledge, and indicate the lexical relation of interest. [26] use Hearst's concept of lexicosyntactic patterns to extract idioms that consist of fixed patterns between two nouns. Basically, their technique works by finding patterns such as "thrills and spills", whose reversals (such as "spills and thrills") are never encountered.

While many idioms do have these properties, many idioms fall on the continuum from being compositional to being partly unanalyzable to completely non-compositional [5]. [8,16], among others, notice that type-based approaches do not work on expressions that can be interpreted idiomatically or literally depending on the context and thus, an approach that considers tokens in context is more appropriate for the task of idiom recognition. A number of token-based approaches have been discussed in the literature, both supervised (Katz and Giesbrecht 2006), weakly supervised [1] and unsupervised [8,24]. [24] present a graph-based model for representing the lexical cohesion of a discourse. Nodes represent tokens in the discourse, which are connected by edges whose value is determined by a semantic relatedness function. They experiment with two different approaches to semantic relatedness: (1) Dependency vectors, as described in [20]; (2) Normalized Google Distance [4]. [24] show that this method works better for larger contexts (greater than five paragraphs). [16] assume that literal and figurative data are generated by two different Gaussians, literal and non-literal and the detection is done by comparing which Gaussian model has a higher probability to generate a specific instance. The approach assumes that the target expressions are already known and the goal is to determine whether this expression is literal or figurative in a particular context. The important insight of this method is that figurative language in general exhibits less semantic cohesive ties with the context than literal language. Their results are inconclusive, due to the small size of the test corpus. [21] investigate the bag of words *topic* representation and incorporate an additional hypothesis–contexts in which idioms occur are more affective. Still, they treat idioms as semantic outliers.

[9] describe several approaches to automatic idiom identification. One of them is idiom recognition as outlier detection. They apply principal component analysis for outlier detection – an approach that does not rely on costly annotated training data and is not limited to a specific type of a syntactic construction, and is generally language independent.

7 Conclusions

In this paper we described an original algorithm for automatic classification of idiomatic and literal expressions. We also compared our algorithm against several competing idiom detection algorithms discussed in the literature. The performance results show that our algorithm generally outperforms [8] model. Note that $CTX+$ is a supervised version of [8], in that the training data here is the true "gold-standard," while in [8] is noisy. A research direction is to incorporate affect into our model. Idioms are typically used to imply a certain evaluation or affective stance toward the things they denote [19,22]. We usually do not use idioms to describe neutral situations, such as buying tickets or reading a book. Similarly to [21] we are exploring ways to incorporate affect into our idiom detection algorithm. Even though our method was tested on verb-noun constructions, it is independent of syntactic structure and can be applied to any idiom type. Unlike [8] approach, for example, our algorithm is language-independent and does not rely on POS taggers and syntactic parsers, which are often unavailable for resource-poor languages. Our next step is to expand this method and use it for idiom discovery. The move will imply an extra step – extracting multiword expressions first and then determining their status as literal or idiomatic.

Acknowledgements. This material is based upon work supported by the National Science Foundation under Grant No. 1319846.

References

1. Birke, J., Sarkar, A.: A clustering approach to the nearly unsupervised recognition of nonliteral language. In: Proceedings of the 11th Conference of the European Chapter of the Association for Computational Linguistics (EACL 2006), Trento, pp. 329–336 (2006)
2. Burnard, L.: The British National Corpus Users Reference Guide. Oxford University Computing Services, Oxford (2000)
3. Cacciari, C.: The place of idioms in a literal and metaphorical world. In: Cacciari, C., Tabossi, P. (eds.) Idioms: Processing, Structure, and Interpretation, pp. 27–53. Lawrence Erlbaum Associates, Hillsdale (1993)

4. Cilibrasi, R., Vitányi, P.M.B.: The Google similarity distance. IEEE Trans. Knowl. Data Eng. **19**(3), 370–383 (2007)
5. Cook, P., Fazly, A., Stevenson, S.: Pulling their weight: exploiting syntactic forms for the automatic identification of idiomatic expressions in context. In: Proceedings of the ACL 2007 Workshop on A Broader Perspective on Multiword Expressions, pp. 41–48 (2007)
6. Cook, P., Fazly, A., Stevenson, S.: The VNC-tokens dataset. In: Proceedings of the LREC Workshop: Towards a Shared Task for Multiword Expressions (MWE 2008), Marrakech, June 2008
7. Cowie, A.P., Mackin, R., McCaig, I.R.: Oxford Dictionary of Current Idiomatic English, vol. 2. Oxford University Press, Oxford (1983)
8. Fazly, A., Cook, P., Stevenson, S.: Unsupervised type and token identification of idiomatic expressions. Comput. Linguist. **35**(1), 61–103 (2009)
9. Feldman, A., Peng, J.: Automatic detection of idiomatic clauses. In: Gelbukh, A. (ed.) CICLing 2013. LNCS, vol. 7816, pp. 435–446. Springer, Heidelberg (2013). doi:10.1007/978-3-642-37247-6_35
10. Fellbaum, C., Geyken, A., Herold, A., Koerner, F., Neumann, G.: Corpus-based studies of German idioms and light verbs. Int. J. Lexicogr. **19**(4), 349–360 (2006)
11. Firth, J.R.: A synopsis of linguistic theory, 1930–1955 (1957)
12. Fukunaga, K.: Introduction to Statistical Pattern Recognition. Academic Press, New York (1990)
13. Glucksberg, S.: Idiom meanings and allusional content. In: Cacciari, C., Tabossi, P. (eds.) Idioms: Processing, Structure, and Interpretation, pp. 3–26. Lawrence Erlbaum Associates, Hillsdale (1993)
14. Hearst, M.A.: Automatic acquisition of hyponyms from large text corpora. In: Proceedings of the 14th Conference on Computational Linguistics (COLING 1992), vol. 2, pp. 539–545. Association for Computational Linguistics, Stroudsburg (1992). http://dx.doi.org/10.3115/992133.992154
15. Katz, G., Giesbrecht, E.: Automatic identification of non-compositional multiword expressions using latent semantic analysis. In: Proceedings of the ACL/COLING-06 Workshop on Multiword Expressions: Identifying and Exploiting Underlying Properties, pp. 12–19 (2006)
16. Li, L., Sporleder, C.: Using Gaussian mixture models to detect figurative language in context. In: Proceedings of the NAACL/HLT 2010 (2010)
17. Mikolov, T., Chen, K., Corrado, G., Dean, J.: Efficient estimation of word representations in vector space. In: Proceedings of Workshop at ICLR (2013)
18. Mikolov, T., Sutskever, I., Chen, K., Corrado, G., Dean, J.: Distributed representations of words and phrases and their compositionality. In: Proceedings of the NIPS (2013)
19. Nunberg, G., Sag, I.A., Wasow, T.: Idioms. Language **70**(3), 491–538 (1994)
20. Pado, S., Lapata, M.: Dependency-based construction of semantic space models. Comput. Linguist. **33**(2), 161–199 (2007)
21. Peng, J., Feldman, A., Vylomova, E.: Classifying idiomatic and literal expressions using topic models and intensity of emotions. In: Proceedings of the 2014 Conference on Empirical Methods in Natural Language Processing (EMNLP), pp. 2019–2027. Association for Computational Linguistics, Doha, October 2014. http://www.aclweb.org/anthology/D14-1216
22. Sag, I.A., Baldwin, T., Bond, F., Copestake, A., Flickinger, D.: Multiword expressions: a pain in the neck for NLP. In: Proceedings of the 3rd International Conference on Intelligence Text Processing and Computational Linguistics (CICLing 2002), Mexico City, pp. 1–15 (2002)

23. Seaton, M., Macaulay, A. (eds.): Collins COBUILD Idioms Dictionary, 2nd edn. HarperCollins Publishers (2002)
24. Sporleder, C., Li, L.: Unsupervised recognition of literal and non-literal use of idiomatic expressions. In: Proceedings of the 12th Conference of the European Chapter of the Association for Computational Linguistics (EACL 2009), pp. 754–762. Association for Computational Linguistics, Morristown (2009)
25. Villavicencio, A., Copestake, A., Waldron, B., Lambeau, F.: Lexical encoding of MWEs. In: Proceedings of the Second ACL Workshop on Multiword Expressions: Integrating Processing, Barcelona, pp. 80–87 (2004)
26. Widdows, D., Dorow, B.: Automatic extraction of idioms using graph analysis and asymmetric lexicosyntactic patterns. In: Proceedings of the ACL-SIGLEX Workshop on Deep Lexical Acquisition (DeepLA 2005), pp. 48–56. Association for Computational Linguistics, Stroudsburg (2005). http://dl.acm.org/citation.cfm?id=1631850.1631856

A Text Mining-Based Framework
for Constructing an RDF-Compliant
Biodiversity Knowledge Repository

Riza Batista-Navarro, Chrysoula Zerva, Nhung T.H. Nguyen,
and Sophia Ananiadou[✉]

School of Computer Science, University of Manchester, Manchester M13 9PL, UK
{riza.batista,chrysoula.zerva,nhung.nguyen,
sophia.ananiadou}@manchester.ac.uk

Abstract. In our aim to make the information encapsulated by biodiversity literature more accessible and searchable, we have developed a text mining-based framework for automatically transforming text into a structured knowledge repository. A text mining workflow employing information extraction techniques, i.e., named entity recognition and relation extraction, was implemented in the Argo platform and was subsequently applied on biodiversity literature to extract structured information. The resulting annotations were stored in a repository following the emerging Open Annotation standard, thus promoting interoperability with external applications. Accessible as a SPARQL endpoint, the repository facilitates knowledge discovery over a huge amount of biodiversity literature by retrieving annotations matching user-specified queries. We present some use cases to illustrate the types of queries that the knowledge repository currently accommodates.

1 Introduction

Big data—huge data collections—are proliferating in many disciplines at a rate that is much faster than what our analytical abilities can handle. One particular discipline that has amassed big data is biological diversity, more popularly known as biodiversity: the study of variability amongst all life forms. This discipline plays a central role in our daily lives, given its implications on ecological resilience, food security, species and subspecies endangerment, and natural sustainability. On the one hand, researchers in this domain collect primary data pertaining to the occurrence or distribution of species, and store this information in a structured format (e.g., spreadsheets, database tables). On the other hand, findings or observations resulting from their analysis of primary data are usually reported in literature (e.g., monographs, books, journal articles or reports), often referred to as secondary data. Written in natural language, secondary data lacks the structure that primary data comes with, rendering the knowledge it contains obscured and inaccessible. In order to make information from secondary data available in a structured and thus searchable form, we have developed a

© Springer International Publishing AG 2017
J.A. Lossio-Ventura and H. Alatrista-Salas (Eds.): SIMBig 2015/2016, CCIS 656, pp. 30–42, 2017.
DOI: 10.1007/978-3-319-55209-5_3

repository containing information automatically extracted from biodiversity literature by a customisable text mining workflow. To maximise its interoperability with external tools or services, we have made the knowledge repository available as a Resource Description Framework (RDF) triple store that conforms with the Open Annotation standard [16]. By means of a few exemplar use cases, we then demonstrate how the repository, accessible as a SPARQL endpoint, facilitates query-based search, thus making the information contained in biodiversity literature discoverable.

A handful of other tools for storing biodiversity information in RDF format exist. Most of them, however, do not have the capability to automatically understand text written in natural language. Tools such as RDF123 [9] and BiSciCol Triplifier [17], for example, accept only data that is already in the form of structured tables. The browser extension Spotter [13] generates RDF-formatted annotations over blog posts, not by automatically extracting information from the textual content but rather by requiring its users to manually enter structured descriptive metadata. Most similar to our work is a system for automatically extracting RDF triples pertaining to species' morphological characteristics, from the literature on Flora of North America [8]. Their semantic annotation application provided the user with an opportunity to revise automatically generated annotations, an option that can also be enabled in our approach. We note though that our work has the additional advantage of being uniquely underpinned by highly customisable and extensible workflows. In this way, when domain experts call for other types of information to be captured, our framework will require only minimal development time and effort to fulfill the task.

2   Methodology

In this section, we present in detail our framework for constructing the knowledge repository. We begin by briefly describing the corpus of biodiversity documents that was utilised, and then outline the various steps in the text mining workflow. Results of evaluating the workflow's performance against gold standard data are then presented. We finally proceed to explaining how the Open Annotation specification was adopted in order to store the information extracted by the text mining workflow from our corpus.

2.1   Document Selection

The Biodiversity Heritage Library (BHL)[1] is a database of biodiversity literature maintained by a consortium of natural history and botanical libraries all over the world. A product of the various partners' digitisation efforts, BHL currently contains almost 110,000 titles, equivalent to almost 50 million pages of text resulting from the application of optical character recognition (OCR) tools on scanned images of legacy materials. For this work, we decided to narrow down

---

[1] http://www.biodiversitylibrary.org.

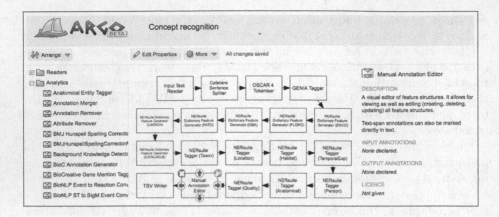

Fig. 1. Argo's workflow construction interface

the scope of the knowledge repository to the requirements of a project called "Conserving Philippine Biodiversity by Understanding Big Data" (COPIOUS). Its aim is to comprehensively collect and store into one knowledge repository, both primary and secondary information on biodiversity in the Philippines. For the purpose of populating the repository with information derived from secondary data, the project requires the development of text mining methods that can extract fine-grained details from biodiversity literature.

To this end, we retrieved the subset of English BHL pages which are relevant to the Philippines. Specifically, we took the union of (1) the set of pages which mention either "Philippines" or "Philippine" within their content, and (2) the set of pages contained by books or volumes whose titles mention "Philippines" or "Philippine". This resulted in a corpus of a total of 155,635 pages, amounting to around 12 GB of data.

2.2   Development of Text Mining Workflow

One of the primary interests of our collaborators in the COPIOUS project is the discovery of fundamental species-centric knowledge, particularly information on species' geographic locations, habitat, anatomical parts as well as authorities (i.e., persons who described them). Guided by user requirements, we cast this work as an information extraction task necessitating: (1) named entity recognition (NER) for six semantic types, namely, taxon, location, habitat, anatomical part, person and temporal expression, and (2) binary relation extraction focussing on the following types of associations: taxon-location, taxon-habitat, taxon-anatomical part and taxon-person.

To carry out these tasks on our corpus, we integrated various natural language processing (NLP) tools into one workflow using the Argo platform [15]. Argo[2] is a web-based, graphical workbench that facilitates the construction and

---

[2] http://argo.nactem.ac.uk.

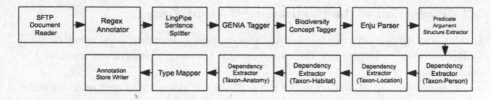

Fig. 2. Text mining workflow

execution of bespoke modular text mining workflows. Underpinning it is a library of diverse elementary NLP components, each of which performs a specific task. Argo's graphical block diagramming interface for workflow construction, shown in Fig. 1, provides access to the component library (at the left-hand side), representing them as blocks that can be interconnected to define processing sequence. Each component can be configured by means of user-selected parameter values, which can be specified via a spanner button that appears upon hovering the mouse pointer on each block.

The workflow that we developed as part of this work, depicted in Fig. 2, combines several components for pre-processing, syntactic and semantic analyses. It begins with an SFTP (Secure File Transfer Protocol) Document Reader which loads the plain-text corpus from a remote server. This is followed by a Regex Annotator which attempts to detect paragraph boundaries based on the occurrence of newline characters. The paragraphs are then segmented by the LingPipe Sentence Splitter [3] into sentences, each of which is decomposed into tokens by the GENIA Tagger [18] which also performs part-of-speech tagging, lemmatisation and chunking. The next component, the Biodiversity Concept Tagger, is a machine learning-based named entity recogniser built upon the NERsuite package [4] that applies a conditional random fields (CRF) model [10] to recognise names under any of the following categories: taxon, location, habitat, anatomical part, temporal expression and person. In Sect. 2.3, we describe in detail the features that we employed in training and applying our CRF model, and present results of the evaluation that we carried out to assess NER performance.

The succeeding components in the workflow contribute towards the relation extraction task. Enju Parser [11] performs deep syntactic parsing and extracts syntactic dependencies amongst sentence tokens. Its outputs are used by the next component, the Predicate Argument Structure Extractor, to compute semantic dependencies in the form of predicate-argument structures. Each of the five instances of the Dependency Extractor component then makes use of the predicate-argument structures to detect relationships between names categorised under the specified entity types. The first instance, for example, detects only relationships between taxon and person names, while the last one captures related taxa and anatomical parts. The Type Mapper [14] ensures that all of the named entities and relations extracted conform with the same annotation schema before they are all saved in Open Annotation format by the last component, the Annotation Store Writer.

2.3   Evaluation of the Biodiversity Concept Tagger

In this section we provide further details on how the conditional random field (CRF) model underpinning the Biodiversity Concept Recogniser, learned the NER task.

The recognition of biodiversity-relevant concept names was cast as a sequence labelling problem, i.e., the assignment of one of `begin`, `inside` and `outside` (BIO) labels to a sentence's tokens. Each token was encoded as a rich set of lexical, orthographic and semantic features, such as:

1. two, three and four-character n-grams
2. token, POS tag and lemma unigrams and bigrams within a window of 3
3. presence of digits or special characters
4. token containing only uppercase letters
5. word shape, with all of token's uppercase letters converted to 'A', lowercase letters to 'a', digits to '0' and special characters to '_'
6. matches against semantically relevant dictionaries

The last type of features were generated based on the Catalogue of Life [6] and the following ontologies: the Gazetteer [2], Environment [7], Uber Anatomy [12], Phenotype and Trait [5], and Flora Phenotype [1] Ontologies.

Our chosen implementation of CRFs is an in-house Java wrapper for the NERsuite package [4] which comes with modules for training new models and applying those models on documents. For training and evaluating our CRF model, we utilised a small collection of BHL documents, annotated in-house as part of the COPIOUS project. This development data set consists of 55 abstract-sized documents in which a total of 2,507 names falling under the categories of taxon, location, habitat, anatomical part, person and temporal expression have been manually annotated according to a set of carefully defined guidelines[3].

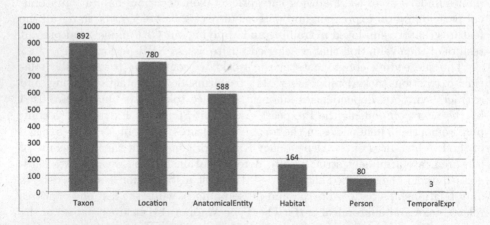

Fig. 3. Frequency distribution of named entities in the development data set

---

[3] http://wiki.miningbiodiversity.org/doku.php?id=guidelines.

Table 1. Named entity recognition performance in terms of micro-averaged precision, recall and F-score

Entity type	TPs	FPs	FNs	Precision	Recall	F-score
Taxon	416	97	476	81.092	46.637	59.217
Location	546	111	234	83.105	70.000	75.992
Habitat	76	24	88	76.000	46.341	57.576
Person	4	4	76	50.000	5.000	9.091
Anatomical part	285	73	303	79.609	48.469	60.254
Temporal expression	0	0	3	100.000	0.000	0.000
Overall	1327	309	1180	81.112	52.932	64.060

Shown in Fig. 3 is the frequency distribution of the gold standard named entity annotations according to type.

Ten-fold cross validation was carried out to evaluate the performance of our CRF-based named entity recogniser. The predicted named entities in the test set of each of the 10 folds were accumulated, yielding the number of true positives (TPs), false positives (FPs) and false negatives (FNs) presented in Table 1. This allowed us to calculate the micro-averaged precision, recall and F-score for each entity type, and the overall F-score of 64.060% (precision = 81.112%, recall = 52.932%). We note that while the overall performance can be considered acceptable, poor results were obtained for certain entity types, i.e., person and temporal expression. This can be attributed to the very sparse named entity annotations for these two types, as observed from Fig. 3. By annotating more entities under these types, we can potentially boost NER performance in future versions of our text mining workflow.

2.4   Adopting the Open Annotation Model

We now describe how our extracted annotations were encoded according to the Open Annotation (OA) specification. The OA Core Data Model is an emerging W3C-recommended standard for encoding associations between any annotation and resource (i.e., what is being annotated). Built upon the Resource Description Framework (RDF), the OA model represents an annotation as having a *body* and a *target*, with the former somehow describing the latter, e.g., by assigning a label or identifier. Following this fundamental idea and other relevant recommendations given in the specification[4], we represented the named entity and relation annotations extracted by our text mining workflow in OA format, as depicted in Fig. 4. We note that for brevity, prefixes were used in this figure instead of full namespaces, e.g., oa for http://www.w3.org/ns/oa#. A relation annotation, represented in the figure as the root of the graph, consists of a body and a target. Since a relation annotation is 'about' a pair of entities, its target is a composite consisting of two named entity annotations. Its body, which

---

[4] http://www.openannotation.org/spec/core.

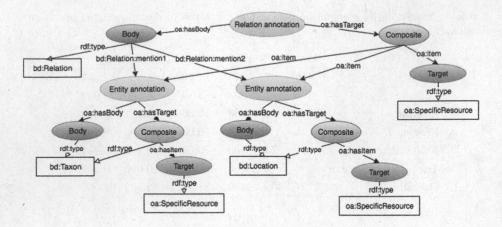

Fig. 4. Our Open Annotation representation of related entities

holds details of the annotation, points to the same pair of entity annotations, but also specifies which of the two serves as `mention1` and `mention2`. Each of the entity annotations, in turn, has a body and a target. In this case, the body node indicates the semantic type, e.g., `Taxon`, `Location`. Meanwhile, the target node points to a composite encapsulating a `SpecificResource`, which stores the exact position of the annotated piece of text within the original document, in terms of character offsets.

Once the RDF triples had been generated, they were automatically loaded onto a new instance of Apache Jena TDB[5], an RDF store which serves as our knowledge repository's backend. This is then exposed as an HTTP-accessible SPARQL endpoint by Apache Jena Fuseki[6].

3   Use Cases

We present some examples of how knowledge discovery can be facilitated by our repository, which can be accessed via HTTP[7] as a SPARQL endpoint that accepts POST requests. The body of the request should be set to a valid SPARQL query while the headers should be configured to hold the following name-value pairs:

- `Accept: text/csv`
- `Content-Type: application/sparql-query`

In Fig. 5, we provide a screenshot of the Postman REST client[8] to show the configuration we used in interfacing with the SPARQL endpoint.

---

[5] https://jena.apache.org/documentation/tdb.
[6] https://jena.apache.org/documentation/fuseki2.
[7] http://nactem.ac.uk/copious-demo/annotations/sparql.
[8] https://www.getpostman.com.

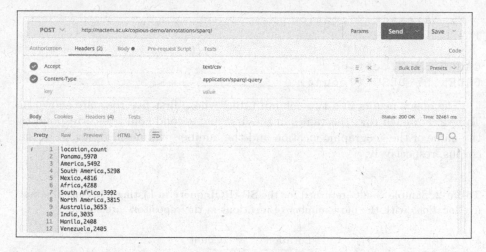

Fig. 5. The Postman REST client with configurations for interfacing with the SPARQL endpoint. The SPARQL query itself is entered in a text area displayed in the Body tab (not shown in the figure).

As a first example, we formulate a query to determine which geographical locations have been catalogued in the repository, and how often they were mentioned in the literature. Shown in Listing 1.1 is a query in SPARQL, the query language for RDF, that retrieves a list of all locations, as well as the number of times that each one was mentioned in the BHL corpus that formed the basis of the repository.

Listing 1.1. A simple SPARQL query example that will retrieve all location names in the repository, ordered by the number of times they are mentioned in our BIIL corpus.

```
PREFIX rdfs: <http://www.w3.org/2000/01/rdf-schema#>
PREFIX oa: <http://www.w3.org/ns/oa#>
PREFIX rdf: <http://www.w3.org/1999/02/22-rdf-syntax-ns
    #>
PREFIX bd: <http://nactem.ac.uk/schema/uima/typesystem/
    MiningBiodiversityTypeSystem#uk.ac.nactem.uima.
    biodiv.>

SELECT  ?location (COUNT(?location) as ?count)

WHERE {
  ?annotation oa:hasBody ?body .
  GRAPH ?body {
    ?target rdf:type bd:Location.
  }
  ?target oa:hasSelector ?selector .
  ?selector oa:default ?x .
```

```
    ?x oa:exact ?location.
}
```

```
GROUP BY ?location
ORDER BY DESC (?count)
```

The query results in a two-column table, whose first five rows are shown in Table 2. For each row, the values in the first and second columns correspond to the name of the geographic location and the number of times it occurred in the corpus, respectively.

Table 2. Sample results returned for the SPARQL query in Listing 1.1. Only the first five locations with the most number of mentions in the repository are shown.

Location	Count
Panama	5970
America	5492
South America	5298
Mexico	4816
Africa	4288

One might also be interested in retrieving the specific documents in the corpus that mentioned a particular name. In our second example, shown in Listing 1.2, file names of BHL documents mentioning "Panama" are retrieved. We note that a regular expression (regex) was employed in matching the word "Panama" against named entities in the knowledge repository. The "i" flag is specified in the regex pattern to enable case-insensitive matching. Since we made use of the string anchors '^' and '$' which respectively match the start and end of named entities, longer strings containing "Panama", e.g., "Panama river", "Panama City", are not returned.

Listing 1.2. An example of a SPARQL query that will retrieve the specific BHL documents mentioning a specific location, in this case, Panama.

```
PREFIX rdfs: <http://www.w3.org/2000/01/rdf-schema#>
PREFIX oa: <http://www.w3.org/ns/oa#>
PREFIX rdf: <http://www.w3.org/1999/02/22-rdf-syntax-ns
    #>
PREFIX bd: <http://nactem.ac.uk/schema/uima/typesystem/
    MiningBiodiversityTypeSystem#uk.ac.nactem.uima.
    biodiv.>
```

```
SELECT  ?location ?source
```

```
WHERE {
```

```
?annotation oa:hasBody ?body .
GRAPH ?body {
  ?target rdf:type bd:Location.
}
?target oa:hasSelector ?selector .
?selector oa:default ?x .
?x oa:exact ?location .
FILTER(regex(?location,"^Panama$", "i")).
?target oa:hasSource ?source .
}

GROUP BY ?location ?source
```

Sample results are presented in Table 3. The file names shown in the second column contain the following document metadata that should allow a user to locate the specific document from BHL: item ID, language, part ID (if available), page ID and year of publication.

Table 3. Sample results returned for the SPARQL query in Listing 1.2. Only the first five rows are shown.

Location	Source document
Panama	22416_English_75050_3163339_1963.txt
Panama	99484_English_31885532_1963.txt
Panama	99484_English_31886231_1963.txt
Panama	71197_English_21871802_unknown.txt
Panama	100797_English_41026_32373656_1928.txt

A user might also be interested in learning which specific geographic locations have been described in the literature as having associations with certain species, e.g., the bird family of hornbills. Shown in Listing 1.3 is a query that retrieves a list of all such locations, ordered by the number of times that the relationship was mentioned in the source document.

Listing 1.3. An example SPARQL query that will retrieve locations related to hornbills, as well as the respective source documents and corresponding number of times the relationship was mentioned.

```
PREFIX rdfs: <http://www.w3.org/2000/01/rdf-schema#>
PREFIX oa: <http://www.w3.org/ns/oa#>
PREFIX rdf: <http://www.w3.org/1999/02/22-rdf-syntax-ns
  #>
PREFIX bd: <http://nactem.ac.uk/schema/uima/typesystem/
  MiningBiodiversityTypeSystem#uk.ac.nactem.uima.
  biodiv.>
```

```
SELECT ?taxon ?loc ?source (COUNT(?loc) as ?count)

WHERE {
  ?annotation oa:hasBody ?body .
  GRAPH ?body {
    ?a rdf:type bd:Relation .
    ?a bd:Relation:mention1 ?mention1 .
    ?a bd:Relation:mention2 ?mention2 .
  }
  ?mention1 oa:hasTarget ?target1 .
  GRAPH ?comp1 {
    ?target1 rdf:type bd:Taxon .
  }
  ?target1 oa:hasSelector ?selector1 .
  ?selector1 oa:default ?default1 .
  ?default1 oa:exact ?taxon .
  FILTER(regex(?taxon, "Hornbill", "i")) .
  ?mention2 oa:hasTarget ?target2 .
  GRAPH ?comp2 {
    ?target2 rdf:type bd:Location .
  }
  ?target2 oa:hasSelector ?selector2 .
  ?selector2 oa:default ?default2 .
  ?default2 oa:exact ?loc .
  ?target2 oa:hasSource ?source .
}

GROUP BY ?taxon ?loc ?source
ORDER BY DESC (?count)
```

In the sample results returned by the above query (Table 4), the second column contains the names of geographic locations which were identified by our text mining workflow as having an association with hornbills. The third and fourth columns respectively indicate the file name of the source document, and the number of times that the location name was described as being related to the species within the document.

Table 4. First five rows returned by the query in Listing 1.3.

Taxon	Location	Source document	Count
Hornbill	Malabar	97438_English_31113063_2001.txt	3
Hornbill	Manilla	97438_English_31113064_2001.txt	3
Hornbill	Sumba	97438_English_31113062_2001.txt	2
hornbills	Africa	133732_English_42321121_unknown.txt	1
Hornbills	Africa	31141_English_6938785_1961.txt	1

4   Conclusion

In this paper, we presented a framework for building a knowledge repository that: (1) applies a customisable text mining workflow to extract information in the form of named entities and relationships between them; (2) stores the automatically extracted knowledge as RDF triples compliant with the Open Annotation specification; and (3) facilitates the discovery of otherwise obscured knowledge by enabling query-based retrieval of annotations from a SPARQL endpoint. We note that the triple store can be exposed via other application programming interfaces, i.e., web services that abstract away from SPARQL to make querying straightforward for non-technical users. We envision that our knowledge repository will facilitate the enhancement of search applications such as information retrieval systems.

Acknowledgments. We would like to thank Prof. Marilou Nicolas for her valuable inputs. This work is funded by the British Council [172722806 (COPIOUS)], and is partially supported by the Engineering and Physical Sciences Research Council [EP/1038099/1 (CDT)].

References

1. Flora Phenotype Ontology. https://bioportal.bioontology.org/ontologies/FLOPO. Accessed 20 Sep 2016
2. Gazetteer. http://bioportal.bioontology.org/ontologies/GAZ. Accessed 20 Sep 2016
3. LingPipe. http://alias-i.com/lingpipe/. Accessed 20 Sep 2016
4. NERsuite: a named entity recognition toolkit. http://nersuite.nlplab.org/. Accessed 20 Sep 2016
5. Plant Trait Ontology. http://www.obofoundry.org/ontology/to.html. Accessed 20 Sep 2016
6. Species 2000 & ITIS Catalogue of Life. Digital resource, September 2016. www.catalogueoflife.org/col. Accessed 20 Sep 2016
7. Buttigieg, P.L., Morrison, N., Smith, B., Mungall, C.J., Lewis, S.E.: The environment ontology: contextualising biological and biomedical entities. J. Biomed. Semant. **4**(1), 43 (2013)
8. Cui, H., Jiang, K., Sanyal, P.P.: From text to RDF triple store: an application for biodiversity literature. In: Proceedings of the Association for Information Science and Technology (ASIST 2010) (2010)
9. Han, L., Finin, T., Parr, C., Sachs, J., Joshi, A.: RDF123: from spreadsheets to RDF. In: Sheth, A., Staab, S., Dean, M., Paolucci, M., Maynard, D., Finin, T., Thirunarayan, K. (eds.) ISWC 2008. LNCS, vol. 5318, pp. 451–466. Springer, Heidelberg (2008). doi:10.1007/978-3-540-88564-1_29
10. Lafferty, J.D., McCallum, A., Pereira, F.C.N.: Conditional random fields: probabilistic models for segmenting and labeling sequence data. In: Proceedings of the Eighteenth International Conference on Machine Learning, pp. 282–289. Morgan Kaufmann Publishers Inc., San Francisco (2001)
11. Miyao, Y., Tsujii, J.: Feature forest models for probabilistic HPSG parsing. Comput. Linguist. **34**(1), 35–80 (2008)

12. Mungall, C.J., Torniai, C., Gkoutos, G.V., Lewis, S.E., Haendel, M.A.: Uberon, an integrative multi-species anatomy ontology. Genome Biol. **13**(1), R5 (2012)
13. Parr, C., Sachs, J., Han, L., Wang, T.: RDF123 and spotter: tools for generating OWL and RDF for biodiversity data in spreadsheets and unstructured text. In: Proceedings of Biodiversity Information Standards Annual Conference (TDWG 2007) (2007)
14. Rak, R., Rowley, A., Carter, J., Batista-Navarro, R., Ananiadou, S.: Interoperability and customisation of annotation schemata in argo. In: Proceedings of the Ninth International Conference on Language Resources and Evaluation (LREC'14), pp. 3837–3842. European Language Resources Association (ELRA), May 2014
15. Rak, R., Rowley, A., Black, W., Ananiadou, S.: Argo: an integrative, interactive, text mining-based workbench supporting curation. Database **2012**, bas010 (2012)
16. Sanderson, R., Ciccarese, P., Van de Sompel, H.: Designing the w3c open annotation data model. In: Proceedings of the 5th Annual ACM Web Science Conference (WebSci 2013), pp. 366–375. ACM, New York (2013)
17. Stucky, B.J., Deck, J., Conlin, T., Ziemba, L., Cellinese, N., Guralnick, R.: The BiSciCol triplifier: bringing biodiversity data to the semantic web. BMC Bioinform. **15**(1), 1–9 (2014)
18. Tsuruoka, Y., Tateishi, Y., Kim, J.-D., Ohta, T., McNaught, J., Ananiadou, S., Tsujii, J.: Developing a robust part-of-speech tagger for biomedical text. In: Bozanis, P., Houstis, E.N. (eds.) PCI 2005. LNCS, vol. 3746, pp. 382–392. Springer, Heidelberg (2005). doi:10.1007/11573036_36

Network Sampling Based on Centrality Measures for Relational Classification

Lilian Berton[2]([✉]), Didier A. Vega-Oliveros[1], Jorge Valverde-Rebaza[1],
Andre Tavares da Silva[2], and Alneu de Andrade Lopes[1]

[1] Department of Computer Science, ICMC, University of São Paulo,
São Carlos, SP 13560-970, Brazil
{davo,jvalverr,alneu}@icmc.usp.br
[2] Technological Sciences Center,
University of Santa Catarina State, Joinville, SC 89219-710, Brazil
lilian.2as@gmail.com, andre.silva@udesc.br

Abstract. Many real-world networks, such as the Internet, social networks, biological networks, and others, are massive in size, which impairs their processing and analysis. To cope with this, the network size could be reduced without losing relevant information. In this paper, we extend a work that proposed a sampling method based on the following centrality measures: degree, k-core, clustering, eccentricity and structural holes. For our experiments, we remove 30% and 50% of the vertices and their edges from the original network. After, we evaluate our proposal on six real-world networks on relational classification task using eight different classifiers. Classification results achieved on sampled graphs generated from our proposal are similar to those obtained on the entire graphs. The execution time for learning step of the classifier is shorter on the sampled graph compared to the entire graph and random sampling. In most cases, the original graph was reduced by up to 50% of its initial number of edges without losing topological properties.

Keywords: Network sampling · Relational classification · Centrality measures · Missing data · Complex networks

1 Introduction

Network data analysis is a topic continuously growing and explored since a lot of real systems can be represented via network structure, such as social, biological and technological domains. The research topic encompasses community detection [5], information propagation [22], disease spreading [15], link prediction [21] and relational classification [10–13]. These studies allow understanding patterns of linking among entities, the relationships between people and systems, the way information spread and how networks are organized [14].

Nevertheless, many of real networks are massive in size, being difficult to be studied in their entirety. In some cases, the network is not entirely available, is complicated to be collected, or even if we have the complete graph, it can

J.A. Lossio-Ventura and H. Alatrista-Salas (Eds.): SIMBig 2015/2016, CCIS 656, pp. 43–56, 2017.
DOI: 10.1007/978-3-319-55209-5_4

be very expensive to run the algorithms on it. So, one possible way to deal with these difficulties is network sampling, i.e., to select a subset of vertices and edges from the entire graph and to study how the sampling processes impact the performance of network applications [1,2,9].

A network can be represented by a set $G = (V, E, W)$, where $V = \{v_1, v_2, \ldots, v_N\}$ is the N vertices, $E = \{e_{ij}\}$ is the M edges and W is the matrix of weights, where w_{ij} denotes the weight of e_{ij}, i.e., the weight of the edge between v_i and v_j. The set V can represent users in a social network, and the set E can describe the friendship among these users. The weights can be defined according to the task, for example, the strength of the relation between two users. A sampled graph $G' = (V', E')$ is a subset from G, such that $V' \cup V$ and $E' \cup E$. The sampled graph G' should preserve important properties of the original network for the addressed task.

For many real-world domains (*e.g.* online social networks, geographic maps, World Wide Web, etc.) networks data do not fit in memory, so, different strategies for sampling have been proposed aiming to reduce the number of vertices or edges of the network. These algorithms can be separated into two groups [17]: (1) The "one-phase sampling algorithms" which select nodes or edges uniformly at random or using some kinds of graph traversal procedure, for example, RDS [7], RWS [18], SBS [6]. These approaches are simple, have low computational cost and low accuracy; (2) The "two-phase sampling algorithms" which construct sampled networks using a graph traversal procedure and some pre- or post-processing step, for example, SPS [16], DPL [23], IFFST-PR [20], Forest Fire Sampling (FFS) [9], *Snowball sampling* using Breadth-First Search (BFS) [8]. These approaches have a high computational cost and better accuracy.

The authors in [19] computed the original and the sampled networks statistics, *i.e.* calculate the representativeness of the sampled subgraph structure comparing it with to the entire network and evaluate the impact of the missing data in the sampled network. In [17] was proposed a generalization of six network sampling algorithms for sampling weighted networks. The evaluation carried out by using network statistics. The approach in [2] is focused in network sampling for relational classification. The authors evaluated the performance of four sampling methods (three random sampling and FFS) by comparing classification accuracy and graph properties (degree, path length, and clustering coefficient). The authors only applied the weighted-vote relational neighbor (wvrn) [13] as the base classifier on three datasets. Here, we perform an extensive research focused on the study of the impact of using sampled networks in relational classification. We exploit a lot of classifiers and datasets with weighted networks. We show that random sampling can take more time in the learning phase and presents significant standard deviation regards accuracy.

This work explores centrality measures for network sampling. The centrality measures indicate how much a vertex is important in some scope. For example, *degree centrality* means how popular is a vertex, *clustering coefficient* and *k-core* means how connected is a vertex, *eccentricity* and *structural holes* means how central a vertex is in the network. We consider these measures (degree (DG), k-core (KC), clustering coefficient (CT), eccentricity (EC) and structural holes (HO)) because they can be calculated in part of the graph and have low computational cost. We select some percentage of vertices with highest and lowest values of these measures and their correspondent edges, then, we remove these vertices and edges from the entire network. Finally, the reduced networks are applied in relational classification. We show the relational classification accuracy obtained by the proposed sampling strategy is as good as the classification accuracy obtained on entire graphs and takes less time in the learning phase.

This paper is an extension of a previous work [3]. The main differences between the two versions are: (1) the addition of two relational classifiers in the experiments, in total we have eight classifiers that work on graphs with known labels for some vertices to predict the labels of the remaining vertices. The classifiers considered are: weighted vote relational neighbor (wvrn), network-only Bayes (no-Bayes), probabilistic relational neighbor (prn), class-distribution relational neighbor (cdrn-norm-cos) and network-only link-based (no-lb). For the network-only link-based classifier we employed models modelink (no-lb-mode), count-link (no-lb-count), binary-link (nolb-binary) and class-distribution-link (no-lb-distrib); (2) the analysis of random sampling where some vertices are randomly chosen to be removed. We show random sampling take more time to build a classification model and lead to higher standard deviation in the classification accuracy; (3) the calculation of two standard evaluation measures (bias score and Pearson's coefficient) to compare how different are the centrality measures of sampled graphs compared to the original graph.

The remaining of this paper is organized as follows. Section 2 presents the proposed approach for network sampling. Section 3 presents the experimental evaluation which analyzes the impact of sampling on relational classification and the network topology. Finally, Sect. 4 presents the conclusions and future works.

2 Proposal

Our proposal consists in an intuitive approach based on exploring the centrality measures of a network to remove some vertices and edges trying to conserve the topology equivalence between the sampled and the full network. Thus, our proposal generates a sample G' from G, i.e. $G' = \sigma(G)$, where σ is a function for sampling G' from G. It is important to note that G' is a sub-graph from G, so $V' \subset V$ and $E' \subset E$. The size of the sample is related to the graph size. We aim to obtain a sample from G in such way it does not affect the performance of the learning task for classification.

The proposed approach is illustrated in Fig. 1. It includes the following steps: (i) calculate a particular centrality measure for all vertices of the network, in this paper we employed the degree (DG), K-core (KC), clustering coefficient (CT), eccentricity (EC) and structural holes (HO) measures [14]; (ii) select some percentage of vertices with the highest (H) or lowest (L) centrality values, in this paper we experiment selecting 30% and 50% of vertices; (iii) remove all selected

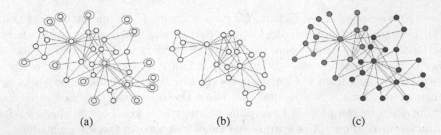

Fig. 1. Proposed approach: (a) Select some % of vertices with lowest (red) or highest (green) centrality measure values from original graph. (b) Remove all the selected vertices and all their edges to obtain the sampled network. In this case, we remove vertices with lowest centrality measure values. (c) Use some learning task on the sampled network, for instance, the relational classification. (Color figure online)

vertices and all their corresponding edges from G, obtaining G'. The sampled graph, G', should be equivalent to the original graph regard the classification task. So distinct classifiers should have a similar performance in both. A relational classifier R takes part of G' as input for learning the classification model. Then, R classifies the remainder of G and evaluates the performance based on the accuracy of the predicted class labels.

All measures employed for sampling can be calculated considering only a fraction of the graph, in a direct way or by applying statistical methods. The measures DG, HO and CT, for instance, can be calculated for each vertex directly. In the case of EC and KC, there are very precise approaches that consider only the vertex community (part of the network). These measures have low computational cost to be calculated and can be applied to large networks.

3   Experimental Results

We carried out a comprehensive evaluation analyzing the quality of sampled graphs generated by different configurations of our proposal compared to the original graph. We used six real-world networks and eight relational classifiers on the original and sampled graphs. We performed two types of evaluations, Sect. 3.2 shows the classification accuracy results, and Sect. 3.3 shows the topological analysis of the sampled and original graphs.

3.1   Data Sets and Experimental Setup

We considered six benchmark data sets[1], which represent real networks and are described in Table 1. Also, we considered that all networks are undirected.

A subgraph G' was sampled from the entire graph G using the centrality measures presented and considering 30% and 50% of vertices with smallest and highest centralities values. For each sample size, we performed 10-fold cross-validation

---

[1] http://netkit-srl.sourceforge.net/data.html.

Table 1. Data sets description.

Datasets	$|V|$	$|E|$	# Classes	$\langle k \rangle$
Cora	4240	35912	7	17.84
Cornell	351	1393	6	3.98
Imdb	1441	51481	2	66.99
Industry	2189	6531	12	10.65
Texas	338	1002	6	3.44
Washington	434	1941	6	4.41

and employed the following relational classifiers: weighted vote relational neigh-bor (wvrn), network-only Bayes (no-Bayes), probabilistic relational neighbor (prn), class-distribution relational neighbor (cdrn-norm-cos) and network-only link-based (no-lb) classifiers, in their Netkit-SRL implementations with standard configurations. For the network-only link-based classifier we employed models modelink (no-lb-mode), count-link (no-lb-count), binary-link (nolb-binary) and class-distribution-link (no-lb-distrib). The area under the ROC curve (AUC) was used as evaluation measure to compare the accuracy of the entire graph G and the sampled graph G'.

3.2 Impact of Sampling on Classification Accuracy

The classification results for the entire graph, random sampling and 30% and 50% of sampled networks are shown in Figs. 2 and 3 respectively, with the accuracy for the six datasets (y-axis) and the ten sampling strategies proposed (x-axis). For the sampling strategies that eliminate 30% of data, the data sets Cora and Imdb achieved the highest accuracy. The Cora dataset also achieved the highest accuracy for sampling strategies that eliminate 50% of vertices.

From the classification results, the Nemenyi post-hoc test [4] was executed to verify the possibility of detecting statistical differences among the sampling strategies. The results for 30% and 50% of sampled networks are shown in Figs. 4 and 5 respectively. On the top of the diagrams is the critical difference (CD) and in the axis are plotted the average ranks of the evaluated techniques, where the lowest (best) ranks are on the left side. When the methods analyzed have no significant difference, they are connected by a black line in the diagram.

According to the Nemenyi statistics, the critical value for comparing the average-ranking of two different algorithms considering the sampling strategy that removes 30% of vertices (Fig. 4) or 50% of vertices (Fig. 5) at 95 percentile in all classifiers (no-Bayes, prn, cdrn-norm-cos, nolb-binary, no-lb-count, no-lb-distrib, no-lb-mode, wvrn) is 6.16.

In all the classifiers there are some sampling strategies that have no statistical difference with the entire graph. It is the case of CT-30H EC-30H, DG-30L, KC-30L, HO-30H and EC-30L for 30% of vertices sampled, and DG-50L, KC-50L, EC-50H, HO-50H, CT-50L and EC-50L for 50% of vertices sampled. Regarding

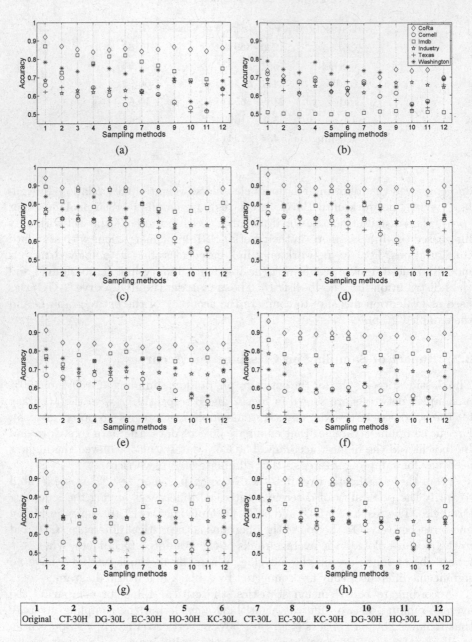

1	2	3	4	5	6	7	8	9	10	11	12
Original	CT-30H	DG-30L	EC-30H	HO-30H	KC-30L	CT-30L	EC-30L	KC-30H	DG-30H	HO-30L	RAND

Fig. 2. Classification results for the entire graph (Original), random sampling and sampling strategies that removes 30% of vertices for the following relational classifiers: (a) no-Bayes; (b) nolb-binary; (c) no-lb-count; (d) no-lb-distrib; (e) no-lb-mode; (f) wvrn; (g) prn; (h) cdrn-norm-cos.

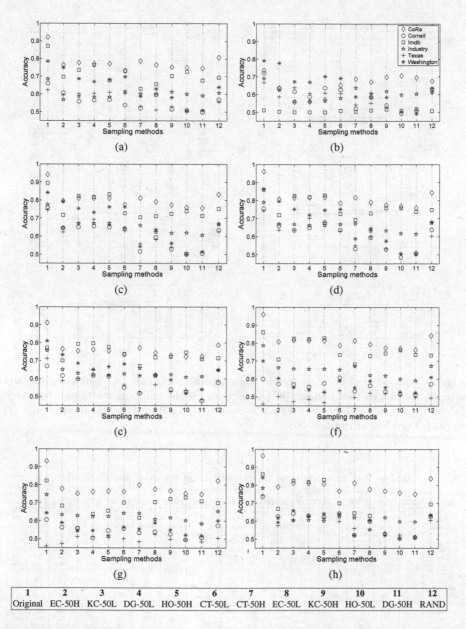

Fig. 3. Classification results for the entire graph (Original), random sampling and sampling strategies that removes 50% of vertices for the following relational classifiers: (a) no-Bayes; (b) nolb-binary; (c) no-lb-count; (d) no-lb-distrib; (e) no-lb-mode; (f) wvrn; (g) prn; (h) cdrn-norm-cos.

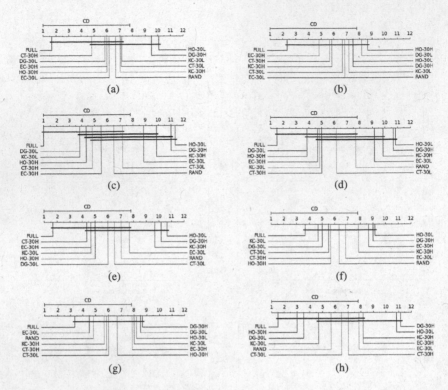

Fig. 4. Nemenyi post-hoc test for the entire graph (original), random sampling and sampling strategies that removes 30% of vertices for the following relational classifiers: (a) no-Bayes; (b) nolb-binary; (c) no-lb-count; (d) no-lb-distrib; (e) no-lb-mode; (f) wvrn; (g) prn; (h) cdrn-norm-cos.

Table 2. Execution time for classification training models (ms).

Classifiers	Dataset	CT-30H	EC-30H	HO-30H	KC-30H	DG-30H	FULL	RAND	Dataset	CT-30H	EC-30H	HO-30H	KC-30H	DG-30H	FULL	RAND
prn		5686	5269	5143	5702	**5572**	6831	5729		5010	4724	4899	4192	**3973**	7980	4712
no-Bayes		756	824	960	552	**511**	1395	748		529	635	744	314	**290**	2841	851
nolb-binary		13498	13398	13857	13380	**13313**	14547	13941		23140	**21688**	22061	26874	26101	23196	**21333**
no-lb-count	CoRa	**12213**	12585	13897	12386	12765	13759	13054	Industry	**23751**	23934	27333	25806	26370	27111	26930
no-lb-distrib		**14218**	14310	14569	14452	14401	15396	14774		27827	26984	**26721**	29931	28834	28148	26904
no-lb-mode		13991	13837	14097	13304	**12924**	17516	13857		27854	25516	**24045**	28688	27881	31027	25543
wvrn		552	622	713	403	**378**	1032	520		281	323	346	187	**180**	541	305
cdrn-norm-cos		18507	18612	17816	18613	19293	**6978**	18505		11052	24836	26513	11761	14414	**3152**	9221
prn		367	380	392	282	**266**	763	370		349	956	776	**270**	493	842	1668
no-Bayes		74	52	52	50	**39**	228	87		54	47	52	44	**40**	206	66
nolb-binary		724	775	789	694	**694**	936	864		809	730	703	680	**601**	914	712
no-lb-count	Cornell	718	809	754	697	**667**	1250	828	Texas	826	773	765	703	**605**	1125	911
no-lb-distrib		753	759	772	**690**	719	918	804		766	681	701	680	**633**	895	742
no-lb-mode		729	703	726	**690**	712	2018	780		657	642	666	674	**664**	2283	735
wvrn		38	49	45	**31**	36	145	69		42	38	42	**37**	40	153	52
cdrn-norm-cos		710	1534	1161	18613	2387	**335**	938		1457	3176	1589	2524	3447	**1312**	1638
prn		943	784	588	595	**573**	2588	978		488	409	449	**390**	681	1088	469
no-Bayes		381	327	558	214	**175**	1042	370		68	72	66	45	**42**	277	121
nolb-binary		959	810	1260	659	**587**	1538	928		967	963	950	888	**804**	1016	970
no-lb-count	Imdb	998	843	1258	693	**605**	1786	1003	Washington	1016	977	957	899	**868**	1124	1145
no-lb-distrib		947	868	1232	662	**595**	1484	1019		1043	955	945	880	**794**	1190	1030
no-lb-mode		970	830	1086	621	**600**	2720	1025		870	948	877	897	**834**	2300	971
wvrn		482	387	695	253	**214**	1050	439		51	54	80	**44**	46	218	76
cdrn-norm-cos		5439	5525	5231	6006	**4716**	5432	6081		2318	1550	3765	3042	3560	**469**	1931

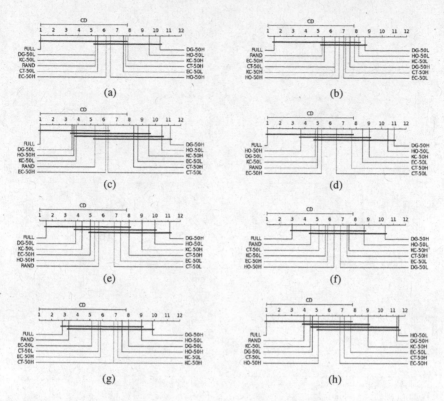

Fig. 5. Nemenyi post-hoc test for the entire graph (Original), random sampling and sampling strategies that removes 50% of vertices for the following relational classifiers: (a) no-Bayes; (b) nolb-binary; (c) no-lb-count; (d) no-lb-distrib; (e) no-lb-mode; (f) wvrn; (g) prn; (h) cdrn-norm-cos.

accuracy, this result indicates that the centrality measures considered, for all the analyzed parameters, were robust and suitable as a sampling strategy.

Table 2 shows the time comparison for the learning step for all classifiers and datasets. We notice that all sampling strategies proposed, considering 30% of sampling, achieve small-time compared to the entire graph and the random approach, especially the strategy DG and KC.

Figure 6 shows the standard deviation to apply a random sampling. On each box, the central mark is the median, the edges of the box are the 25th and 75th percentiles, the whiskers extend to the most extreme data points not considered outliers and the outliers are plotted individually. We noticed that for all datasets (except Cora) and classifiers there is a considerable higher dispersion, indicating that a random choice can result in lower accuracy.

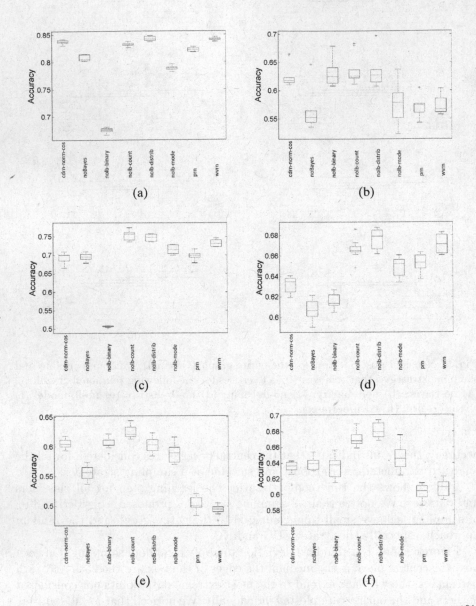

Fig. 6. Boxplots showing the dispersion for ten random sampling when removing 50% of vertices to the following datasets: (a) Cora; (b) Cornell; (c) Imdb; (d) Industry; (e) Texas; (f) Washington.

3.3 Impact of Sampling on Network Topology

We analyze the impact of the sampling methods by comparing the structures of the original and the sampled networks. In Table 3, we have the fraction of remaining edges after applying the sampling methods, according to the target vertices (with highest (H) or lowest (L) centrality value) and removal percentage (30 or 50%). The bold values highlight the techniques and parameters that achieve similar accuracy results to the entire network, i.e., with no significant difference for all the classifiers. We observed that the EC and HO measures are inversely proportional to the final fraction of remaining edges. This occurs because for the EC, the most central vertices have the lowest values and for the HO measure, hubs tend to have larger ego-networks; ergo, the centrality values are lower.

Table 3. Descriptions of edges on sampled graphs.

Dataset	Stat.	#ed-30H	#ed-30L	#ed-50H	#ed-50L	Dataset	Stat.	#ed-30H	#ed-30L	#ed-50H	#ed-50L
CoRa	DG	0.216	0.916	0.077	**0.780**	Industry	DG	0.079	0.952	0.029	**0.883**
	KC	0.266	0.918	0.093	**0.785**		KC	0.090	0.951	0.039	**0.887**
	HO	**0.912**	0.231	**0.767**	0.090		HO	**0.952**	0.094	**0.882**	0.035
	EC	**0.718**	0.389	**0.477**	0.191		EC	**0.758**	0.354	**0.354**	0.152
	CT	**0.555**	**0.600**	0.293	**0.368**		CT	**0.601**	**0.553**	0.314	**0.290**
Cornell	DG	0.067	0.853	0.017	**0.666**	Texas	DG	0.083	0.835	0.026	**0.634**
	KC	0.140	0.849	0.040	**0.670**		KC	0.182	0.829	0.070	**0.643**
	HO	**0.853**	0.080	**0.659**	0.020		HO	**0.835**	0.083	**0.626**	0.034
	EC	**0.662**	0.326	**0.479**	0.114		EC	**0.685**	0.529	**0.492**	0.196
	CT	**0.500**	**0.702**	0.051	**0.553**		CT	**0.474**	**0.720**	0.085	**0.559**
Imdb	DG	0.226	0.881	0.069	**0.648**	Washington	DG	0.082	0.857	0.017	**0.680**
	KC	0.284	0.881	0.086	**0.653**		KC	0.178	0.863	0.074	**0.692**
	HO	**0.872**	0.252	**0.637**	0.096		HO	**0.855**	0.084	**0.685**	0.028
	EC	**0.499**	0.347	**0.264**	0.146		EC	**0.724**	0.268	**0.524**	0.114
	CT	**0.601**	**0.553**	0.314	**0.290**		CT	**0.467**	**0.745**	0.081	**0.616**

We notice that there exist different values of removed edges from the original network, without strongly affecting the accuracy of the classifiers (in bold). This variation of extracted edges, some larger than 50%, suggest that depending on the expected requirements, it can be privileged in the sampling process:

1. The maximal removal of edges by removing a low proportion of vertices.
2. Equivalent removal proportion of edges and vertices.
3. The minimal removal of edges by removing a high proportion of vertices.

In the first case, by removing 30% of vertices we have the sampling method CT-30H. For the third case, we have the methods DG-50L, KC-50L, and HO-50H. The left bold sampling strategies are in the second case.

Notwithstanding reducing the number of vertices and edges from the original network do not statistically affect the classification results, the topological properties are sensibly affected by the removal. For instances, removing 30% of

Table 4. Topological bias regression and Pearson's coefficient results of sampled graphs.

Datasets	Measures	CT-30H	CT-30L	DG-30H	DG-30L	EC-30H	EC-30L	HO-30H	HO-30L	KC-30H	KC-30L	RAND
CoRa	Bias-1	0.756	10.159	8.284	1.062	1.850	9.534	**0.187**	10.399	3.022	1.129	1.937
	Bias-2	4.000	1.000	0.500	7.000	**0.000**	**0.000**	49.000	0.500	0.5	6.000	**0.000**
	Bias-3	0.302	0.353	0.047	0.180	**0.017**	0.038	0.107	0.036	0.097	0.199	0.036
	Pearson	0.999	**1.00**	0.999	0.999	0.999	0.999	0.999	0.999	0.999	0.999	**1.000**
Cornell	Bias-1	0.181	0.035	0.427	0.024	0.473	1.702	0.024	3.345	1.872	**0.003**	0.201
	Bias-2	**0.039**	1.584	0.495	3.475	0.505	0.515	3.475	0.581	0.716	3.521	0.802
	Bias-3	0.985	0.719	1.000	0.492	0.189	0.297	0.492	0.715	0.636	0.514	**0.078**
	Pearson	**0.999**	0.999	-0.127	0.999	0.999	0.999	0.999	-0.153	0.987	0.999	0.999
Imdb	Bias-1	0.111	0.062	0.880	0.453	**0.033**	0.566	0.431	2.220	0.133	0.458	0.169
	Bias-2	0.455	4.102	0.341	4.954	0.409	3.636	4.977	0.352	**0.114**	4.955	0.329
	Bias-3	0.795	**0.607**	0.746	0.683	0.679	0.696	0.698	0.719	0.752	0.684	0.708
	Pearson	0.991	0.995	**0.999**	0.954	0.997	0.999	0.955	0.999	0.999	0.954	0.997
Industry	Bias-1	0.225	0.193	1.371	0.169	0.591	3.296	0.179	3.725	0.264	0.186	**0.089**
	Bias-2	0.000	0.000	0.000	0.000	0.000	0.000	0.000	0.000	0.000	0.000	0.000
	Bias-3	0.925	**0.623**	0.891	0.635	0.763	0.781	0.642	0.846	0.883	0.635	0.766
	Pearson	0.999	0.999	0.988	0.999	0.999	**1.000**	0.999	0.279	0.999	0.999	0.999
Texas	Bias-1	0.123	0.201	1.336	0.073	0.106	0.242	0.073	2.100	0.087	**0.057**	0.166
	Bias-2	0.377	1.321	**0.309**	3.265	0.525	0.969	3.265	**0.309**	0.543	1.722	0.642
	Bias-3	0.935	0.753	1.000	0.539	0.194	**0.045**	0.539	0.802	0.809	0.533	0.172
	Pearson	0.999	0.999	-0.097	0.999	0.999	0.999	0.999	-0.130	0.484	0.999	0.999
Washington	Bias-1	**0.022**	0.217	1.535	0.139	0.382	0.604	0.161	3.577	0.289	0.129	0.172
	Bias-2	**0.113**	0.632	0.889	0.655	0.703	0.926	0.675	0.875	0.959	0.666	0.707
	Bias-3	0.910	0.854	0.741	0.623	0.255	**0.189**	0.463	0.327	0.713	0.636	0.364
	Pearson	0.999	0.999	-0.018	0.999	0.999	0.999	0.999	-0.089	0.862	0.999	0.999

vertices with the highest degree centrality (k_i) produces a more homogeneous distributed network (tending to a Poisson or regular graph) and the average degree decays. On the other hand with the same proportion, removing the least connected vertices produce networks with more heterogeneous degree distribution than the original graph.

To better analyze the impact of sampling in the network topology, we used two standard evaluation measures to compare how different are the centrality measures of sampled graphs with respect to the original graph. Therefore, the first evaluation measure is the *bias score*, which shows how much the sampled network differs from its original regarding a particular topology statistic. This measure is defined as $Bias = \frac{(true-observed)}{true}$, where *true* represents the value of any topology statistic calculated on the original network, while *observed* represents the value of the same topology measure but calculated on the sampled graph. Bias score can be negative (over-estimates) or positive (under-estimates), but we use the absolute value to make them comparable in our analysis [19].

The second evaluation measure used is the Pearson's Coefficient, which measure the correlation between a set of topology statistics calculated from the original network, W, and from the sampled network, Z. Thus, this measure is calculated as $Pearson = \sum_{i=0}^{n}(W - \overline{W})(Z - \overline{Z})/(\sqrt{\sum_{i=0}^{n}(W - \overline{W})^2}\sqrt{\sum_{i=0}^{n}(Z - \overline{Z})^2})$. The coefficient value is between -1 and 1, where -1 means that two networks are negatively linearly correlated and 1 means that they are positively linearly correlated [17].

Table 4 shows the results of bias and Pearson's coefficient for comparison of different topological statistics of original and sampled networks by removing 30% of vertices. Therefore, Bias-1, Bias-2, and Bias-3 show the bias regression results calculated considering assortativity, closeness and clustering coefficient, respectively. To compute the Pearson's coefficient, we employed the following network topology statistics: transitivity coefficient, average path, assortativity, average degree, betweenness, closeness, page rank, eigenvector, k-core, clustering coefficient, and structural holes. Values emphasized (in bold) indicate the best (low) bias regression and (high) Pearson's coefficient values for each dataset.

From Table 4 we observe that, in most of the cases, the bias for all topological statistics considered are relatively low, i.e. the sampled graphs generated from our proposal lose a small ratio of topological information from the original network, especially if we use CT and EC. Also, the sampled graphs have a high and positive linear correlation with the original graph, except when considering highest values of DG and lowest values of HO. Another remark is the fact that, in most of the cases, removing highest (H) values of centrality measures conduct to lose less topological information from the original network that if remove lowest (L) ones.

4   Conclusion

In this paper, we proposed a strategy for network sampling by exploring five centrality measures: DG, KC, CT, EC, HO and eliminating vertices with 30% or 50% of lowest or highest centrality values. All centrality measures considered have a low order of complexity and are computationally applicable in real networks scenarios. Moreover, they can be calculated in part of the graph.

The proposed approach reduces the original graph in 50% or even more and the accuracy results remain statistically similar to the obtained with the entire network, i.e., the impact on classification results obtained by entire networks is minimal when compared with those achieved by sampled networks. We have applied the proposed strategy in six real networks considering eight different relational classifiers. A lot of measures were robust in accuracy for all classifiers and on all networks. Moreover, the execution time for the learning step of the classifiers is smaller in the sampling strategies proposed when compared to the entire graph and random sampling and the loss of network topological information to the original is low.

Acknowledgments. This work was partially supported by the São Paulo Research Foundation (FAPESP) grants: 2013/12191 − 5 and 2015/14228 − 9, National Council for Scientific and Technological Development (CNPq) grants: 140688/2013 − 7 and 302645/2015 − 2, and Coordination for the Improvement of Higher Education Personnel (CAPES), Brazil.

References

1. Ahmed, N.K., Neville, J., Kompella, R.: Network sampling designs for relational classification. In: The 6th International AAAI Conference on Weblogs and Social (2012)
2. Ahmed, N.K., Neville, J., Kompella, T.: Network sampling: from static to streaming graphs. ACM Trans. Knowl. Discov. Data **8**(2), 7:1–7:56 (2013)
3. Berton, L., Vega-Oliveros, D., Valverde-Rebaza, J., Silva, A.T., Lopes, A.: The impact of network sampling on relational classification. In: SIMBig 2016 - SNMAM track. CEUR-WS.org (2016)
4. Demšar, J.: Statistical comparisons of classifiers over multiple data sets. JMLR **7**, 1–30 (2006)
5. Fortunato, S.: Community detection in graphs, CoRR abs/0906.0612v2 (2010)
6. Frank, O.: The Sage Handbook of Social Network Analysis. Sage publications, London (2011)
7. Gile, K.J., Handcock, M.S.: Respondent-driven sampling: an assessment of current methodology. Sociol. Methodol. **1**(40), 285–327 (2010)
8. Lee, S., Kim, P., Jeong, H.: Statistical properties of sampled networks. Phys. Rev. E **73**, 016102 (2006)
9. Leskovec, J., Faloutsos, C.: Sampling from large graphs. In: SIGKDD 2006, pp. 631–636 (2006)
10. Lopes, A.A., Bertini, J.R., Motta, R., Zhao, L.: Classification based on the optimal K-associated network. In: Zhou, J. (ed.) Complex 2009. LNICSSITE, vol. 4, pp. 1167–1177. Springer, Heidelberg (2009). doi:10.1007/978-3-642-02466-5_117
11. Lu, Q., Getoor, L.: Link-based classification. In: ICML 2003, pp. 496–503 (2003)
12. Macskassy, S.A., Provost, F.J.: A simple relational classifier. In: 2nd Workshop on Multi-Relational Data Mining (2003)
13. Macskassy, S.A., Provost, F.J.: Classification in networked data: a toolkit and a univariate case study. JMLR **8**, 935–983 (2007)
14. Newman, M.E.J.: Networks: An Introduction. Oxford University Press, Oxford (2010)
15. Pastor-Satorras, R., Vespignani, A.: Epidemic spreading in scale-free networks. Phys. Rev. Lett. **86**(14), 3200–3203 (2001)
16. Rezvanian, A., Meybodi, M.R.: Sampling social networks using shortest paths. Physica A Stat. Mech. Appl. **424**(C), 254–268 (2015)
17. Rezvanian, A., Meybodi, M.R.: Sampling algorithms for weighted networks. Soc. Netw. Anal. Mining **6**(1), 1–22 (2016)
18. Yon, S., Lee, S., Yook, S.H., Kim, Y.: Statistical properties of sampled networks by random walks. Phys. Rev. E **75**, 46114 (2007)
19. Smith, J.A., Moody, J., Morgan, J.H.: Network sampling coverage II: the effect of non-random missing data on network measurement. Soc. Netw. **48**, 78–99 (2017)
20. Tong, C., Lian, Y., Niu, J., Xie, Z., Zhang, Y.: A novel green algorithm for sampling complex networks. J. Netw. Comput. Appl. **59**, 55–62 (2016)
21. Valverde-Rebaza, J., Valejo, A., Berton, L., Faleiros, T., Lopes, A.: A naïve bayes model based on overlapping groups for link prediction in online social networks. In: ACM SAC 2015, pp. 1136–1141 (2015)
22. Vega-Oliveros, D., Berton, L., Lopes, A., Rodrigues, F.: Influence maximization based on the least influential spreaders. In: SocInf 2015, Co-located with IJCAI 2015, vol. 1398, pp. 3–8 (2015)
23. Yoon, S.-H., Kim, K.-N., Hong, J., Kim, S.-W., Park, S.: A community-based sampling method using DPL for online social networks. Inf. Sci. Int. J. **306**(C), 53–69 (2015)

Dictionary-Based Sentiment Analysis Applied to a Specific Domain

Laura Cruz[1](✉), José Ochoa[1,2](✉), Mathieu Roche[3](✉),
and Pascal Poncelet[4](✉)

[1] Universidad Nacional de San Agustín, Arequipa, Peru
lcruzq@unsa.edu.pe
[2] Universidad Católica San Pablo, Arequipa, Peru
jeochoa@ucsp.edu.pe
[3] TETIS (CIRAD, CNRS, AgroParisTech, Irstea), Paris, France
mathieu.roche@cirad.fr
[4] LIRMM (CNRS, Univ. Montpellier), Montpellier, France
pascal.poncelet@lirmm.fr

Abstract. The web and social media have been growing exponentially
in recent years. We now have access to documents bearing opinions
expressed on a broad range of topics. This constitutes a rich resource
for natural language processing tasks, particularly for sentiment analysis. Nevertheless, sentiment analysis is usually difficult because expressed
sentiments are usually topic-oriented. In this paper, we propose to automatically construct a sentiment dictionary using relevant terms obtained
from web pages for a specific domain. This dictionary is initially built by
querying the web with a combination of opinion terms, as well as terms of
the domain. In order to select only relevant terms we apply two measures
$AcroDef_{MI3}$ and *TrueSkill*. Experiments conducted on different domains
highlight that our automatic approach performs better for specific cases.

Keywords: Text mining · Web mining · Sentiment analysis

1 Introduction

The web and social media have been growing exponentially in recent years, which
constitutes a rich resource for sentiment analysis tasks. For instance, social net-
working sites enable users to express their thoughts and opinions about products
[1] and companies are increasingly taking these opinions into account to make
better decisions [11]. Sentiment analysis currently involves a process to identify the sentiment orientation of opinions. The latter are highly unstructured by
nature, thus requiring the application of Natural Language Processing (NLP)
techniques [17].

Obviously, documents may include opinions about several topics, but terms[1]
used to express opinions are usually specific and highly correlated to a particular domain [6]. For instance, the sentence *"The fruit is organic"* would be

---

[1] In this paper, we use *term* in order to characterize linguistic features.

© Springer International Publishing AG 2017
J.A. Lossio-Ventura and H. Alatrista-Salas (Eds.): SIMBig 2015/2016, CCIS 656, pp. 57–68, 2017.
DOI: 10.1007/978-3-319-55209-5_5

very unusual in movie domain and then irrelevant in this case. However, it is obviously useful in agricultural domain. Both machine learning and dictionary-based approaches have been proposed in the literature to tackle these issues. For instance, a machine learning method applying text categorization techniques was proposed in [12]. By this method, graphs, minimum cut formulation, context, and domain were considered to extract subjective portions of documents.

Some dictionary-based approaches are currently available for general applications (e.g. SentiWordNet[2]). They are not really appropriate for specific domains and new approaches have been developed to automatically learn the dictionary. These methods generally assume that positive (resp. negative) adjectives or verbs appear more frequently near a positive (resp. negative) seed term [9]. For instance, in [16,19], the authors propose an unsupervised learning algorithm for getting a dictionary in order to classify reviews considering seed terms to calculate the semantic orientation of phrases.

In this paper, we propose a new approach to automatically learn expressed opinions. We first focus on a new method for selecting relevant candidate terms from a set of documents. As many candidate terms may be extracted we propose to use two different but complementary measures to select the most representative ones: $AcroDef_{MI3}$ and $TrueSkill$. Furthermore, in order to highlight the fact that of our approach is well useful for extracting terms for a specific domain we compare our proposal to the well-known SentiWordNet.

The paper is organized as follows. Our approach is presented in Sect. 2. The experimental setup is described in Sect. 3. In Sect. 4, we present and discuss the obtained results. Concluding remarks are presented in Sect. 5.

2   Our Approach

The main process of our proposal is depicted in Fig. 1 which involves the following steps:

1. First, a huge corpus for a specific domain is created by querying the web in order to get positive and negative documents relative to this domain.
2. Some pre-processing methods are performed over the documents in order to get the language of the document, remove tags, scripts, images, etc.
3. This step forms the core of the process. It focuses on the selection of terms that could be classified as positive (resp. negative) for the domain. To do, so first, a part-of-speech tagging is performed on documents in order to focus on nouns and adjectives since they are well relevant for extracting opinions[3]. In order to find such terms (adjectives or nouns) as in [9], we follow the following hypothesis: *the closer a positive (resp. negative) adjective/noun to another positive (resp. negative) adjective/noun, the more positive (resp. negative)*

---

[2] http://sentiwordnet.isti.cnr.it.

[3] For simplicity, in this paper, we only report experiments that have been conducted on nouns and adjectives. Other experiments have been done by using adverbs and verbs.

it is. Accordingly, to this, we apply a window size algorithm for selecting the relevant terms closest to given seed terms. Finally, as many candidate terms may be generated, an efficient filtering approach is applied by using *AcroDef$_{MI3}$* and *TrueSkill* to select the most relevant positive and negative terms. Finally based on the results, two lexicons of positive and negative terms are generated.

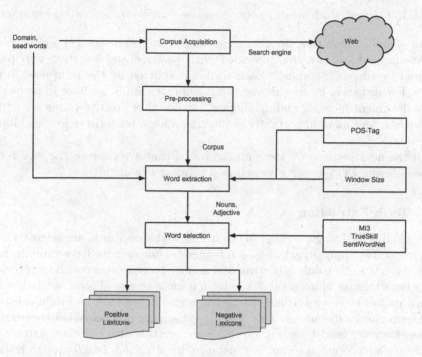

Fig. 1. Lexicons are automatically obtained from web pages for a specific domain filtered by seed terms. Extraction of relevant terms are then obtained by evaluating the relevance of candidate terms that are obtained after the analysis of the documents.

Basically, our approach could be used in many different domains. So in order to highlight its generality, experiments have been conducted on four different domains. They will be described in the experimental section. In the next sections, we describe more in detail the different steps.

2.1 Corpus Acquisition

It is now well admitted that some terms can be positive, neutral, or negative depending of the domain. Nevertheless, some terms are positive or negative irrespective of the domain (e.g. *good*). The main idea of our approach is thus to start the process by considering adjectives which are positive or negative in all domains. These terms will be considered as *seed terms*. We thus select the two following seed sets: P = {*good, nice, excellent, positive, fortunate, correct, superior*}, Q = {*bad, nasty, poor, negative, unfortunate, wrong, inferior*}. From these sets, we ensure a positive (resp. negative) web page retrieved from the other web pages related to a given domain. The following query illustrates an example of what is generated to get only positive documents about Genetic Modified Organism (GMO):

+GMO +good −bad −nasty −poor −negative −unfortunate −wrong −inferior

where + and − mean that the document must have (+) or not (−) a given term.

At the end, we are thus provided with positive and negative web pages denoted by corpus$^+$, corpus$^-$. Each corpus is splitted by the term used in the query. For instance, by considering the previous example we have in corpus$^+$ a set of document focusing mainly on *good*, i.e. no other positive terms are within documents, and more importantly not having a negative term (e.g. *−bad −nasty*, and so on).

In the next section, we focus on the terms that are close to the seed terms by considering POS-Tagging as well as a window size algorithm.

2.2   Term Extraction

First of all HTML tags, scripts, blank spaces and stop words are removed from web pages. We apply a part-of-speech tagger (in our case we have experimented *TreeTagger*[4]) to keep only adjectives and nouns. To be relevant with the previous hypothesis that an opinion candidate term is close to a seed term, a window size algorithm has been used. It aims at finding terms in both left and right sides of a seed term given a distance k. This distance is then the number of left (resp. right) terms of a given seed term. By varying k we are this able to better extract the most correlated opinion terms. For instance, by applying *TreeTagger* to retrieve adjectives (i.e. JJ) and nouns (i.e. NNS) as illustrated in Fig. 2.

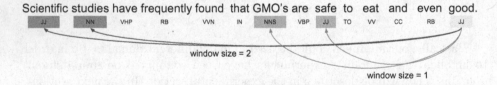

Fig. 2. An example of applying a window size algorithm on the *good* seed term.

In Fig. 2, the '*good*' term is a positive seed term and its nearest adjective is '*safe*' given a $k = 1$ distance. Likewise, '*scientific*' and '*studies*' terms are retrieved with distance of $k = 2$. In addition, '*safe*' is a positive candidate term

---

[4] http://www.cis.uni-muenchen.de/~schmid/tools/TreeTagger/.

because it occurs close to the positive seed term (i.e. *good*). In this sense, we can have a set of opinion terms that can be candidates to be included into the resulting dictionaries.

To get the correlation score and the usefulness on our specific domain (here GMO in the example) of each extracted term two approaches have been used: $AcroDef_{MI3}$ and *TrueSkill* and this is described in the next section.

2.3   Candidate Term Selection

From the set of candidate terms, we thus have to filter the most relevant ones: the positive (resp. negative) terms that are very specific to a domain. In order to select the relevant candidate opinion terms, we propose to adapt the statistical measure $AcroDef_{MI3}$ [14,15] (see Algorithm 1 where we illustrate only for positive terms, the process for the negative terms is similar) as well as a probabilistic measure based on *TrueSkill* [8,10] (see Algorithm 2).

The $AcroDef_{MI3}$ measure: To filter associations extracted at the previous step, we use a ranking function in order to delete the irrelevant adjectives associations placed at the end of a list. Several quality measures in the literature are based on ranking functions. They are brought out of various fields: Association rules extraction [7], terminology extraction [4], and so forth. One of the most commonly used measures to compute a sort of relationship between the terms, called co-occurrence, is Church's Mutual Information (*MI*). The formula is the following [3]:

$$MI(x, y) = \log \frac{nb(x, y)}{nb(x)nb(y)} \tag{1}$$

This measure tends to extract rare and specific co-occurrences according to [4]. The Cubic Mutual Information (*MI3*) is an empirical measure based on *MI* that enhances the impact of frequent co-occurrences. This measure defined by formula (2) gives interesting results [5,18].

$$MI3(x, y) = \log \frac{nb(x, y)^3}{nb(x)nb(y)} \tag{2}$$

Like many other studies based on web resources, the *nb* function used by the *MI* and *MI3* measures represents the number of web pages provided by a search engine.

Our approach relies on the dependence calculation of two terms, i.e. seed terms (*st*), and candidate term (*ct*). This is based on the number of pages given by a search engine with the queries '*st* and *ct*' and '*ct* and *st*'. This dependence is computed in a given domain D (for instance $D = \{GMO\}$). Then we apply $AcroDef_{MI3}$ (formula (3)) described in [14]:

$$AcroDef_{I3}(st, ct) = \log \frac{(nb(st \text{ and } ct \text{ and } D) + nb(ct \text{ and } st \text{ and } D))^3}{nb(st \text{ and } D) \times nb(ct \text{ and } D)} \quad (3)$$

The selection of terms is based on the application of Algorithm 1.

Algorithm 1. Term selection algorithm using $AcroDef_{MI3}$

Require: *corpus, seed terms = P, terms of the domain*
Ensure: correlation score values for each *term*
1: **for** each *corpus* **do**
2:     $terms^+ = window\_size(corpus^+, P)$
3:     **for** term in $terms^+$ **do**
4:         given each seed term and terms of the domain compute the correlation score:
5:             $score \leftarrow max(AcroDef_{MI3})$

The *TrueSkill* measure: Unlike $AcroDef_{MI3}$, in *TrueSkill*, terms are extracted for each positive (resp. negative) page against k random negative (resp. positive) pages, and then the scores for each term are computed. Therefore, after having this outcome *TrueSkill* can give a score for each term of the positive page. This score depends on how many times it appears in the positive page so that it increases or decreases if it is also found on a negative page. The principle of *TrueSkill* is illustrated in Fig. 3.

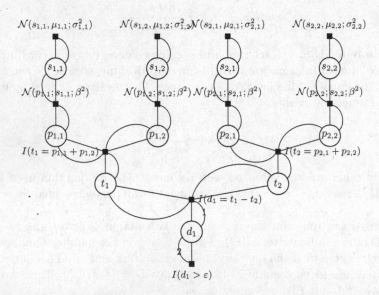

Fig. 3. *TrueSkill* measure provides a score for each term selected given the positive and negative corpora.

In Fig. 3, we have $S = \{s_{1,1}, s_{1,2}, ..., s_{1,n}\}$ and $S = \{s_{2,1}, s_{2,2}, ..., s_{2,n}\}$ where s are the learned score value for each term in positive and negative web pages. p is the performance value for each term, which depends mainly of previously score s of the term; t is the sum of total performance for each term in the corpus. As *TrueSkill* learns s according its outcome of matches, we set a high punctuation for $corpus^+$, and less punctuation for $corpus^-$. This process is detailed in Algorithm 2.

Algorithm 2. Term selection algorithm using *TrueSkill*

Require: *corpus, seed_terms(P, Q)*
Ensure: correlation score values for each *term*
1: $k = 10$ number of matches for each *corpus*.
2: **for** each *corpus* **do**
3:     $terms^+ = window\_size(corpus^+, P)$
4:     **for** k random *corpus*$^-$ **do**
5:         $terms^- = \text{window\_size}(corpus^-, Q)$
6:         given each term compute the correlation score:
7:         $score \leftarrow TrueSkill(terms^+, terms^-, t = [1, 2])$

Finally, *TrueSkill* gives a score to each term of the corpus in a match, and those values are updated for each match. On one hand, if a term is often found in a $corpus^-$ its value tends to decrease. On the other hand, if it is in a $corpus^+$ its value will increase. If the term is found in both corpora it tends to be constant. The velocity of the score increases or decreases depending on the term combination in each corpus and the number of matches.

3   Experiments

In order to evaluate our approach, experiments over four datasets were conducted. First we focused on both measures $AcroDef_{MI3}$ and *TrueSkill* to evaluate their efficiency for pruning candidate terms. Second we evaluated the opinion classification task with both measures. Finally in order to really evaluate our automatic obtained dictionaries, classification are also compared with a general dictionary. *SentiWordnet* is a lexical resource for opinion mining, mainly it comprises 21479 adjectives and 117798 nouns, and assigns three sentiment scores to each word, i.e. positive, negative, and neutral.

3.1   Datasets

In order to show that our approach is generic, we use four datasets on very different domains: agriculture, movie, kitchen, and book:

- On the agricultural domain, tweets have been collected, and have been manually labeled. We obtained a corpus of 183 tweets, i.e. 72 positive and 111 negative tweets.

– Available resources[5] of the movie domain were introduced in [13] with 1000 positive and 1000 negative opinion documents.
– Finally, the kitchen and book domains[6] introduced in [2] have both 1000 positive and 1000 negative opinions.

Table 1 shows the number of candidate terms related to each domain after applying the window size algorithm with $k = 1$, and for each seed term we get the first 20 web pages.

Table 1. Total of inferred lexicon terms by domain.

Lexicon	Agriculture		Movie		Kitchen		Book	
	P	N	P	N	P	N	P	N
Adjective	146	83	104	72	157	26	168	87
Noun	334	207	247	169	335	81	330	197

3.2 Results of $AcroDef_{MI3}$ and $TrueSkill$

In the following WS stands for the terms extracted after the window size algorithm. We thus compared $MI3$: seed terms with WS followed by $AcroDef_{MI3}$ and TS: seed terms with WS followed by $TrueSkill$.

Figure 4 shows the normalized scores of all measures over each term by using the min-max scale algorithm. The window score is based on the frequency of a given term after applying the window size algorithm. As expected we thus have a high number of terms with low score. We can notice that terms have the more distributed score after applying $AcroDef_{MI3}$ and $TrueSkill$.

3.3 Classification

As the context of $SentiWordnet$ is not exactly the same as ours, the neutral class is considered as follows. For a term, we compute the difference between its positive and negative score and if the result is greater than zero we assign the term as positive otherwise as negative.

We have positive and negative lexicons (dictionaries) for each dataset (i.e. agriculture, movie), as shown in Table 1. In order to validate the algorithms we calculate F-$Score$ values for each domain. Figures 5, 6, 7 and 8 show the F-$Score$ values using the built lexicons with our approach and SentiWordNet. On our experiments, the F-$Score$ is evaluated using the lexicons with a score greater than a given threshold.

---

[5] http://www.cs.cornell.edu/People/pabo/movie-review-data/.
[6] https://www.cs.jhu.edu/~mdredze/datasets/sentiment/.

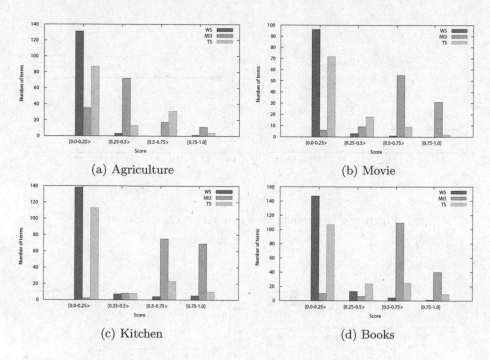

Fig. 4. Lexicons for each domain with their normalized score.

Table 2. Top F-score result of each domain classification, TrueSkill (TS) improves the results of *SentiWordnet* (SWN) in some cases.

	Agriculture		Movie		Kitchen		Book	
	P	N	P	N	P	N	P	N
Approaches	SWN	MI3	SWN	TS	SWN	SWN	SWN	TS
F-Score	0.62	**0.55**	0.66	**0.66**	0.70	0.54	0.67	**0.62**

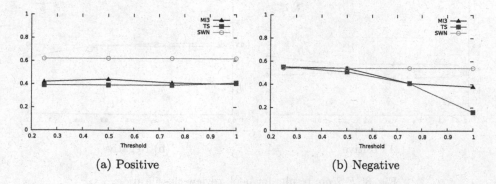

Fig. 5. *F-Score* results for agriculture tweet classification.

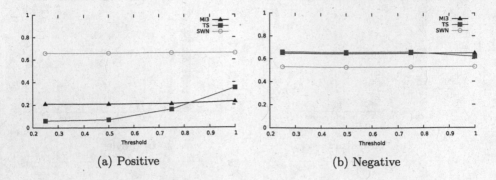

(a) Positive   (b) Negative

Fig. 6. *F-Score* results for movie review classification.

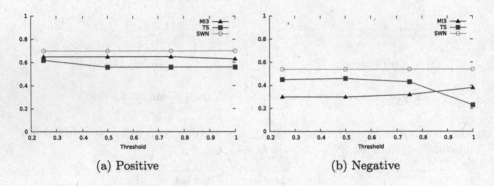

(a) Positive   (b) Negative

Fig. 7. *F-Score* results for kitchen review classification.

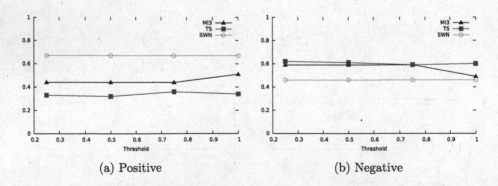

(a) Positive   (b) Negative

Fig. 8. *F-Score* results for book review classification.

4   Discussion of the Results

In order to evaluate *F-Score* results, Table 2 shows the high values obtained for each dataset when the inferred lexicons for each domain are considered. For predicting negative elements, the *F-Score* values of *TrueSkill* are 0.66 and 0.62 for movie and book domains respectively, and 0.55 for agriculture domain based on $AcroDef_{MI3}$.

To sum up, our approach performs better with *F-Score* results than Senti-WordNet for negative reviews. However, when positive reviews are considered, SentiWordNet performs better.

Figure 4 shows that kitchen domain is more generic than other domains due to the high number of terms (≈ 70) with a high score $[0.75-1.0]$ obtained with $AcroDef_{MI3}$. This could explain that SentiWordNet performs better than *TrueSkill* and $AcroDef_{MI3}$ for positive and negative reviews for this domain.

5   Conclusion

In this paper, we proposed a dictionary-based algorithm for sentiment analysis. Our approaches used $AcroDef_{MI3}$ and *TrueSkill* to compute an association score between each term and its sentiment orientation (i.e. positive, negative). The extraction of these new terms related to each domain is obtained using the window size algorithm. This enables us to automatically create dictionaries that have been proved useful to identify positive and negative documents of specific domains.

In future work, we plan to extend our approach to other languages (e.g. French and Spanish), and we would like to study the behavior of our methods with other domains by using multi-word terms in our lexicons.

Acknowledgement. This work has been supported and funded by FONDECYT and SONGES project (http://textmining.biz/Projects/Songes) (FEDER and Occitanie).

References

1. Amine, A., Hamou, R.M., Simonet, M.: Detecting opinions in tweets. Int. J. Data Min. Emerg. Technol. **3**(1), 23–32 (2013)
2. Blitzer, J., Dredze, M., Pereira, F.: Biographies, bollywood, boomboxes and blenders: domain adaptation for sentiment classification. In: Proceedings of the 45th Annual Meeting of the Association for Computational Linguistics (ACL 2007), pp. 187–205 (2007)
3. Church, K.W., Hanks, P.: Word association norms, mutual information, and lexicography. Comput. Linguist. **16**, 22–29 (1990)
4. Daille, B.: Study, implementation of combined techniques for automatic extraction of terminology. In: Klavans, J.L., Resnik, P. (eds.) The Balancing Act: Combining Statistical and Symbolic Approaches to Language, pp. 49–66. MIT Press, Cambridge (1996)

5. Downey, D., Broadhead, M., Etzioni, O.: Locating complex named entities in web text. In: Proceedings of the 20th International Joint Conference on Artificial Intelligence (IJCAI 2007), pp. 2733–2739 (2007)

6. Duthil, B., Trousset, F., Roche, M., Dray, G., Plantié, M., Montmain, J., Poncelet, P.: Locating complex named entities in web text. In: Proceedings of the 22nd International Conference on Database and Expert Systems Applications (DEXA 2011), pp. 457–465 (2007)

7. Guillet, F., Hamilton, H.J.: Quality Measures in Data Mining. Springer, Heidelberg (2007)

8. Guo, S., Sanner, S., Graepel, T., Buntine, W.: Score-based Bayesian skill learning. In: Flach, P.A., Bie, T., Cristianini, N. (eds.) ECML PKDD 2012. LNCS (LNAI), vol. 7523, pp. 106–121. Springer, Heidelberg (2012). doi:10.1007/978-3-642-33460-3_12

9. Harb, A., Plantie, M., Dray, G., Roche, M., Trousset, F., Poncelet, P.: Web opinion mining: how to extract opinions from blogs? In: Proceedings of the 5th International Conference on Soft Computing as Transdisciplinary Science and Technology (CSTST 2008), pp. 211–217 (2008)

10. Herbrich, R., Minka, T., Graepel, T.: TrueSkill(TM): a Bayesian skill rating system. In: Advances in Neural Information Processing Systems, vol. 20, pp. 569–576. MIT Press (2007)

11. Marrese-Taylor, E., Velásquez, J.D., Bravo-Marquez, F., Matsuo, Y.: Identifying customer preferences about tourism products using an aspect-based opinion mining approach. Procedia Comput. Sci. **22**, 182–191 (2013)

12. Pang, B., Lee, L.: A sentimental education: Sentiment analysis using subjectivity summarization based on minimum cuts. In: Proceedings of the ACL, pp. 271–278 (2004)

13. Pang, B., Lee, L., Vaithyanathan, S.: Thumbs up?: sentiment classification using machine learning techniques. In: Proceedings of the ACL 2002 Conference on Empirical Methods in Natural Language Processing, EMNLP 2002, vol. 10, pp. 79–86. Association for Computational Linguistics, Stroudsburg (2002)

14. Roche, M., Prince, V.: *AcroDef*: a quality measure for discriminating expansions of ambiguous acronyms. In: Kokinov, B., Richardson, D.C., Roth-Berghofer, T.R., Vieu, L. (eds.) CONTEXT 2007. LNCS (LNAI), vol. 4635, pp. 411–424. Springer, Heidelberg (2007). doi:10.1007/978-3-540-74255-5_31

15. Roche, M., Prince, V.: A web-mining approach to disambiguate biomedical acronym expansions. Informatica (Slovenia) **34**(2), 243–253 (2010)

16. Turney, P.D.: Thumbs up or thumbs down?: semantic orientation applied to unsupervised classification of reviews. In: Proceedings of the 40th Annual Meeting on Association for Computational Linguistics, ACL 2002, pp. 417–424. Association for Computational Linguistics, Stroudsburg (2002)

17. Varghese, R., Jayasree, M.: Aspect based sentiment analysis using support vector machine classifier. In: 2013 International Conference on Advances in Computing, Communications and Informatics (ICACCI), pp. 1581–1586, August 2013

18. Vivaldi, J., Màrquez, L., Rodríguez, H.: Improving term extraction by system combination using boosting. In: Raedt, L., Flach, P. (eds.) ECML 2001. LNCS (LNAI), vol. 2167, pp. 515–526. Springer, Heidelberg (2001). doi:10.1007/3-540-44795-4_44

19. Wang, G., Araki, K.: Modifying so-PMI for Japanese weblog opinion mining by using a balancing factor and detecting neutral expressions. In: Proceedings of the Conference of the North American Chapter of the Association for Computational Linguistics (NAACL-Short 2007), pp. 189–192 (2007)

A Clustering Optimization Approach
for Disaster Relief Delivery:
A Case Study in Lima-Perú

Jorge Vargas-Florez, Rosario Medina-Rodríguez[(✉)], and Rafael Alva-Cabrera

Department of Engineering, Pontificia Universidad Católica del Perú, Lima, Peru
r.medinar@pucp.pe

Abstract. During the last decade, funds to face humanitarian operations have increased approximately ten times. According to the Global Humanitarian Assistance Report, in 2013 the humanitarian funding requirement was by US\$ 22 billion, which represents 27.2% more than the requested in 2012. Furthermore, the transportation cost represents between one-third to two-thirds from the total logistics cost. Therefore, a frequent problem in a disaster relief is to reduce the transportation cost by keeping an adequate distribution service. The latter depends on a reliable delivery route design, which is not easy to do considering a post-disaster environment, where the infrastructures and sources could be inexistent, unavailable or inoperative. This paper tackles this problem, regarding the constraints, to deliver relief aids in a post-disaster state (like an eight-degree earthquake) in the capital of Perú. The routes found by the hierarchical ascending clustering approach, solved with a heuristic model, achieved a sufficient and satisfactory solution.

1 Introduction

Humanitarian response recognizes two phases after a disaster: the life-saving and the life-sustaining actions [16]. The first one consists in carrying activities that aim to preserve life, like removal debris and rescue victims. The second one involves the provision of aid kits and services such as food, water, temporary shelter, medical care, and protection [1]. As mentioned in [5], the initial relief dispatch is more oriented to attend communities promptly, while waiting to be completed the disaster assessment. In [7], the authors, described that under a state of catastrophe, the depot's network is not expected to respond adequately because its storage and distribution capacity loses its simple operability. Moreover, [9] confirms that transport is the second largest general budget of humanitarian organizations, after staff. Thus, planning transportation routes (VRP, Vehicle Routing Problem) is one of the most important problems of combinatorial optimization and is widely studied with many applications in the real world, like distribution logistics and transport [17]. The humanitarian delivery in disasters cases are concerned to optimize; maximizing unsatisfied demand, minimizing travel time a the total delivery delay [2]. There are three basic approaches for

J.A. Lossio-Ventura and H. Alatrista-Salas (Eds.): SIMBig 2015/2016, CCIS 656, pp. 69–80, 2017.
DOI: 10.1007/978-3-319-55209-5_6

modeling the problem: First, (i) the vehicle route is represented by a binary variable of multiple indexes, that define the vehicle and route identification. Then, (ii) the construction of a dynamic network flow model whose outputs are the vehicle and material flows, that have to be parsed to construct vehicle routes and loads. Finally, (iii) to enumerate all feasible routes between all pairs of supply and demand nodes [11]. The latter open a data analysis utilization based on pattern mining, data mining or clustering to optimize delivery routes. For our purposes, we decided to use the last mentioned approach.

In literature, numerous works have been proposed to deal with the problem of spatial clustering on data associated with natural disasters. Early papers attacked the issue of an emergency evacuation, for example, [12] presented a spatial decision support system to emergency planning. Their approach, based on Geographical Information System (GIS) software, evaluated two issues: (i) static, processing the data from a mathematical, statistical and logical point of view and; (ii) dynamic, establish the terrain for evacuation under certain assumptions and with some specified policies. Then, in [4], an ambulance allocation to improve the rescue process of victims, was proposed. The authors used a spatial clustering combined with fuzzy logic to allocate the correct number of ambulances to each spatial objects grouped into a cluster after an earthquake. Later, [15] described a method to evacuate Shin-Hua City, Taiwan, after an earthquake. They analyze the spatial correlation between objects taking into account six indexes associated to route characteristics. Moreover, [11] proposed a hierarchical cluster and route procedure (HOGCR) for coordinating vehicle routing in large-scale post-disaster distribution and evacuation activities. Recently, [10] evaluated the survival Kobe-1995-earthquake manufacturing plants and their post-earthquake economic impact. They used a geographical clustering technique combined with a micro-econometric approach.

In Perú, from 1582 to 2007, occurred 47 earthquakes with magnitudes between 6.0° to 8.6° on the Richter scale. At least ten were greater than 8.0°, and 100% of them took place between the center and the south coast area of Perú; where Lima is located. [8], presented the 7.9° earthquake in Pisco-Perú in 2007, which killed more than 500 people and affected more than 655 thousand people; who demanded water, food, shelter, clothes, etc. The assistance for victims was distributed through multiple civil defense committees, led by the *Instituto de Defensa Civil* (INDECI). Despite the efforts, they could not manage a proper relief delivery. However, the authors stand out the following conclusions: firstly, receiving donations and transport they were improvised. Secondly, humanitarian aid was distributed haphazardly. Thirdly, duplication of supplies in some closer areas, while the most remote ones, received partial support. Fourthly, distribution of inappropriate aid relief and unfit food for consumption (rotten food, expired date drugs, etc.).

From these statements, three points can be highlighted: (i) the high risk that would suffer the Peruvian capital in a potential major earthquake due to its sociodemographic and seismic location. (ii) The failed National delivery relief case described before and; (iii) the importance of supplies transportation to humanitarian operations. Therefore, it is necessary to provide an optimal, efficient and resilient route design regarding the constraints of a post-disaster environment. For this purpose, we propose an approach based on the hierarchical ascending classification that seems the best option considering the expected bad conditions of the roads in a post-disaster environment.

Following this brief introduction and review, the paper is organized as follows: Sect. 2, describes the followed methodology. Then, Sect. 3, exposes the medical aid relief delivery to Lima and Callao in an eventual earthquake, analyzing the Hierarchical Ascendant Classification (HAC) approach for humanitarian distribution. In the end, we conclude with an analysis of the solution with the lowest travel time, and also some future research works are described.

2   Methodology

In order to find an approximate solution to the process of delivery aid relief in Lima, we propose the use of a vehicle routing solution that must be efficient and resilient. The efficiency is related to the ability to provide the intervention with fewer resources, while resilience condition is related to the capacity to ensure the operation on time, even whether infrastructures and sources are inexistent. Therefore, our goal is to identify the routes to be used in the distribution of humanitarian aid. The methodology is composed of the following stages:

1. Obtain the information about the actual Peruvian humanitarian system from government entities (i.e. INDECI).
2. Perform an analysis about cartography in the territory of study, to identify the most vulnerable, exposed and threatened areas.
3. Carry out a research of the available models to solve the vehicle routing problem.
4. Identify the factors that affect directly the time of transporting humanitarian aid.
5. Medical supplies and Central warehouses depots are grouped in clusters taking into account factors obtained in the previous step.
6. Finally, costs are identified following the applied method of Hierarchical Ascendant Classification approach.

The Peruvian humanitarian supply network in this research is composed of Medical Supplies (AM for "Almacén de Insumos Médicos") and Central Warehouses (AC for Almacén Central"); both located in Lima and pre-defined by INDECI. We follow a heuristic called "cluster-first route-second", which determines clusters of customers compatible with vehicle capacity and solves a traveling salesperson problem for each cluster [13]. Thus, for this proposal we apply the Hierarchical Ascending Classification, forming clusters using the total adjusted travel time for each route instead of the Euclidean distance as a proximity measure. In order to make this adjustment, we used as criterion "how critical is the condition of the affected region", which is based on measurements of vulnerability, accessibility, exposure and proximity to each district of Lima, where each AM is located.

3 Results and Discussion

The HAC analysis performed by using dendrograms [18] is an efficient tool for the cluster identification task which combines many features. For instance, it is possible to separate the population into homogeneous clusters (low within-variability and high between variability). In this proposal, we have considered five features that describe vulnerabilities: seismic location; socioeconomic state; access to delivery point; exposure to hazards and; proximity to the central depot (AC).

Step 1: we use an advanced statistical analysis tool (XLSTAT) and the vulnerability scales obtained from the INDECI categorization criteria, as shown in Table 1. Then, based on this criteria, we obtained a summary of the value associated with each type of vulnerability and the district where they belong.

Table 1. INDECI scale values corresponding to the vulnerability type

Socio-economic vulnerability	Night accessibility	Day accessibility	Hazards exposition	Seismic vulnerability
Very low (1)	Very good (1)	Very good (1)	Low impact (1)	Low (1)
Low (2)	Good (2)	Good (2)	Average impact (2)	Relativity high (2)
Average (3)	Regular (3)	Regular (3)	High impact (3)	High (3)
High (4)	Bad (4)	Bad (4)	Very high impact (4)	Very high (4)
Very high (5)	Very bad (5)	Very bad (5)		

However, to apply the HAC approach, it is necessary to standardize our values for the existence of a correlation. For this purpose, we used the method of the "maximum magnitude of 1" [6]. This method divides the value of each variable by its maximum value, obtaining values between 0 and 1. A summary of both values, original and standardized, can be seen in Table 2.

Step 2: we apply the HAC approach on the new standardized data, choosing which Central Warehouses (ACs) will supply store whose Medical Supplies Warehouses (AMs). This choice was based on the shortest distance from AC to AM; for instance, if the distance between AC1 and AM1 is less than the distance between AC2 and AM1, then it will supply AC1. The results are shown in the dendrograms in Fig. 1; where the horizontal dotted lines divide the collection of points in three clusters set by AMs.

Step 3: finally, we apply the algorithm proposed by [3], looking for routes with lower cost, linking each AM to clusters. The final result shows each AM supplied by each AC, as can be seen in Table 3.

Table 2. Summary of vulnerabilities and standardized vulnerabilities, where: (SE: Socio economic; DA: Day accessibility; NA: Night accessibility; HE: Hazard exposition; SZ: Seismic Zoning) for each supply depot and its respective delivery point

Supply depots	Delivery point	Vulnerabilities					Distance AC to AM (Km)	Standardized vulnerabilities and distances					
		SE	DA	NA	HE	SZ		SE	DA	NA	HE	SZ	Distance AC to AM
AC1	AM3	2	4	3	5	2	2.06	1.0	0.8	0.75	1.0	1.0	0.08
AC1	AM4	2	4	3	4	2	1.77	1.0	0.8	0.75	0.8	1.0	0.06
AC1	AM14	2	4	3	5	2	1.04	1.0	0.8	0.75	1.0	1.0	0.04
AC1	AM15	1	4	4	3	1	19.78	0.5	0.8	1.0	0.6	0.5	0.70
AC1	AM16	1	4	4	3	1	19.71	0.5	0.8	1.0	0.6	0.5	0.70
AC1	AM17	1	4	2	3	1	17.44	0.5	0.8	0.5	0.6	0.5	0.62
AC1	AM18	1	4	4	3	2	21.58	0.5	0.8	1.0	0.6	1.0	0.76
AC1	AM19	1	4	4	3	2	24.37	0.5	0.8	1.0	0.6	1.0	0.86
AC1	AM20	1	4	4	3	2	24.87	0.5	0.8	1.0	0.6	1.0	0.88
AC1	AM21	1	4	4	3	2	22.87	0.5	0.8	1.0	0.6	1.0	0.81
AC1	AM22	1	3	2	4	2	5.87	0.5	0.6	0.5	0.8	1.0	0.21
AC1	AM24	2	3	2	4	2	3.07	1.0	0.6	0.5	0.8	1.0	0.11
AC1	AM25	2	4	2	4	2	1.13	1.0	0.8	0.5	0.8	1.0	0.04
AC2	AM1	2	4	1	2	1	6.29	1.0	0.8	0.25	0.4	0.5	0.22
AC2	AM2	2	4	2	5	1	6.06	1.0	0.8	0.5	1.0	0.5	0.21
AC2	AM5	2	5	3	3	1	10.21	1.0	1.0	0.75	0.6	0.5	0.36
AC2	AM6	2	4	2	1	1	2.75	1.0	0.8	0.5	0.2	0.5	0.10
AC2	AM7	2	3	1	2	2	12.54	1.0	0.6	0.25	0.4	1.0	0.44
AC2	AM8	2	5	3	2	1	8.48	1.0	1.0	0.75	0.4	0.5	0.30
AC2	AM9	1	5	4	4	1	28.32	0.5	1.0	1.0	0.8	0.5	1.00
AC2	AM10	2	5	3	1	1	9.93	1.0	1.0	0.75	0.2	0.5	0.35
AC2	AM11	2	3	2	4	2	2.44	1.0	0.6	0.5	0.8	1.0	0.09
AC2	AM12	2	4	3	3	1	1.92	1.0	0.8	0.75	0.6	0.5	0.07
AC2	AM13	2	4	1	3	1	6.43	1.0	0.8	0.25	0.6	0.5	0.23
AC2	AM23	2	3	2	4	2	4.60	1.0	0.6	0.5	0.8	1.0	0.16
AC2	AM26	2	4	3	4	1	0.76	1.0	0.8	0.75	0.8	0.5	0.03
AC2	AM27	2	4	1	2	1	4.96	1.0	0.8	0.25	0.4	0.5	0.18
Maximal value		2	5	4	5	2	**28.32**						

3.1 Distances Evaluation

In this section, we present an analysis of post-disaster distances to be covered by routes. In the beginning, Euclidean ideal distances have been considered however they must be corrected, so we employed the real distance (or linear distance) which takes into account streets and walls. The "Correction Factor' will affect the transportation time, representing a realistic post-disaster condition; for example streets with debris and transport infrastructures collapsed like bridges.

According to the Peruvian Ministry of Transport and Communications, the poor accessibility post-disaster can cause variations up to 30 min, corresponding to 50% of the average transportation time. Also, reviewing historical events recorded by INDECI, it concludes that poor accessibility in affected areas

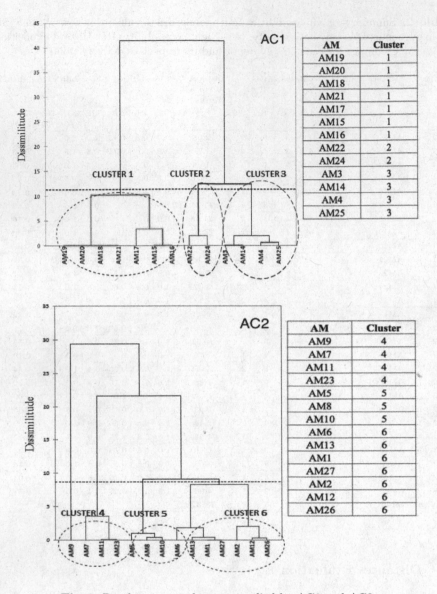

Fig. 1. Dendrograms - clusters supplied by AC1 and AC2.

increases between 25% to 100%, due to collapsed infrastructure or debris. For instance, the transportation time from AM1 will be increased by 65% (due to the correction factor is 165% or 1.65). Thus, because its location has a 4 level accessibility (40%) and also increased by 25%, because its seismic vulnerability corresponds to a level 1 (see Table 4); finally 40% + 25% = 65%.

Table 3. Summary of routes and distances (ideal and real) obtained with HAC, followed by the difference between both distances and its variance, respectively

Clusters	HAC routes									Ideal distance (Km)	Real distance (Km)	Diff. (Km)	Var. (%)
Cluster 1	AC1	AM17	AM21	AM20	AM19	AM18	AM16	AM15	AC1	52.15	58.54	6.39	10.92
Cluster 2	AC1	AM22	AM24	AC1						12.83	16.56	3.73	22.52
Cluster 3	AC1	AM4	AM25	AM3	AM14	AC1				7.63	11.62	3.99	34.34
Cluster 4	AC2	AM9	AM7	AM11	AM23	AC2				68.93	87.37	18.44	21.11
Cluster 5	AC2	AM5	AM8	AM10	AC2					37.35	47.29	9.94	21.02
Cluster 6	AC2	AM12	AM27	AM1	AM13	AM2	AM6	AM26	AC2	21.59	28.94	7.35	25.40
										200.48	**250.32**	**49.84**	**24.86**

3.2 Distribution Expenses Evaluation

Highlighting the difference between ideal and real distance is important. The first one does not consider the location of streets or walls of buildings and it is obtained charting a line between every point (origin and destiny). On the other hand, the real distance takes into account all these points and in cases like this, it could represent a variation until 25% of the total distance traveled (See Table 3). After obtaining the routes and its associated distances, we need to determine the type of transportation that will be used to estimate the required resources. The best vehicle to be employed in the humanitarian distribution due to its capacity and potency is the pick-up (4×4) according to [9], whose key features are described in Table 5.

According to INDECI, each medicine packaging unit or emergency backpack should be able to supply at least two people. It is recommended an average weight of 8 Kg, corresponding to a backpack with a capacity of 20 to 40 L; for practical calculations, we consider an intermediate value of 30 L. Then, considering one vehicle, we can calculate the number of backpacks to carry and how many people they would help. For instance, every trip that makes one transport, will attend 200 people.

$Backpacks$ = volume occupied by medicines in one vehicle; $3\,\mathrm{m}^3 \approx 3000\,\mathrm{L}$.

$$3000\,\mathrm{L} \times \frac{1\,\mathrm{Backpack}}{30\,\mathrm{L}} = 100\,\mathrm{Backpacks}$$

$People$ = attention capacity for backpack $\times$ quantity of backpacks in one pick up

$$\frac{2\,\mathrm{people}}{1\,\mathrm{Backpack}} \times 100\,\mathrm{Backpacks} = 200\,\mathrm{people}$$

Table 4. First, the percentage values to be increased in the distances, according to INDECI. Second, the calculated "Correction Factor" to be applied for each supply depot considering the type of vulnerability.

Supply depot	Bad accessibility (day and night)	Seismic vulnerability	Correction factor
AM1	40%	25%	165%
AM2	40%	25%	165%
AM3	40%	50%	190%
AM4	40%	50%	190%
AM5	50%	25%	175%
AM6	40%	25%	165%
AM7	30%	50%	180%
AM8	50%	25%	175%
AM9	50%	25%	175%
AM10	50%	25%	175%
AM11	30%	50%	180%
AM12	40%	25%	165%
AM13	40%	25%	165%
AM14	40%	50%	190%
AM15	40%	25%	165%
AM16	40%	25%	165%
AM17	40%	25%	165%
AM18	40%	50%	190%
AM19	40%	50%	190%
AM20	40%	50%	190%
AM21	40%	50%	190%
AM22	30%	50%	180%
AM23	30%	50%	180%
AM24	30%	50%	180%
AM25	40%	50%	190%
AM26	40%	25%	165%
AM27	40%	25%	165%
AC1	30%	50%	180%
AC2	40%	25%	165%

Furthermore, we proceeded to group the provinces of Lima and Callao in four main sectors: North-Lima, South-Lima, Lima-Center, East-Lima, and Callao; to estimate the number of affected people to assist (see Table 6). Lima and Callao have 49 districts, and some of them do not have points of medical supplies depots. Meanwhile, other ones have more than one depot. For this reason, we consider that each depot will support the victims by the sector where they belong to

Table 5. Pick-up (4 × 4) features. *Source: Nissan/Toyota; UN Refugee Agency and International Federation of Red Cross and Crescent Societies.*

Feature	Description
Engine power	2500 cc
Number of cylinders	4
Average performance	45 km/gal
Fuel	Diesel
Loading capacity	1 TM cc
Volume occupied by drugs	$3\,m^3$
Drug packaging unit	20–40 L

Table 6. Estimated affected population in the largest Lima's districts in case a major earthquake, according to [14].

District	Affected population	Sector
Ate	69954	East-Lima
Callao	195954	Callao
Carabayllo	127612	North-Lima
Chorrillos	51918	South-Lima
Comas	242235	North-Lima
Lima	18674	Lima-Center
Lurigancho	74186	East-Lima
Lurin	36312	South-Lima
Pachacamac	15260	South-Lima
Puente Piedra	144323	North-Lima
San Juan de Lurigancho	314549	East-Lima
San Juan de Miraflores	128435	South-Lima
Ventanilla	14435	Callao
Villa el Salvador	113993	South-Lima
Villa Maria del Triunfo	133171	South-Lima

(see Table 7); regardless districts which are part of it. The support must be done proportionally to victims' amount in each district.

In Table 8, we indicate the total amount of victims to be supported for each route (where each clustering became one route), considering the real distance. Moreover, we describe the cost of fuel used to support all the victims considering the least covered distance; which is S/. 160,520.62 approximately. We can also provide additional valuable information, like the amount of trips required to support all the victims considering the number of vehicles used.

Table 7. The number of victims that will be supported by each depot. *Source: INEI, SIRAD.*

Sector	Total victims	Depots	Total	Victims to aid for each depot
North-Lima	540 254	AM8 y AM2	2	270 127
South-Lima	483 274	AM7	1	483 274
Center-Lima	64 280	AM1, AM6, AM13 y AM27	4	16 070
East-Lima	490 985	AM5, AM9 y AM10	3	163 662
Callao	216 942	AM3, AM4, AM11, AM12, AM14, AM15, AM16, AM17, AM18, AM19, AM20, AM21, AM22, AM23, AM24, AM25, AM26	17	12 762
	1 795 735			

Table 8. The total amount of victims to be supported for each route, considering corrected distance and the cost of fuel

Clusters	Ideal distance (Km)	Real distance (Km)	Trips	Routes	Victims treated	Fuel cost (S/.)
Cluster 1	52.15	58.54	447	Route 1	89 334	8010.36
Cluster 2	12.83	16.56	128	Route 2	25 524	656.25
Cluster 3	7.63	11.62	256	Route 3	51 048	920.97
Cluster 4	68.93	87.37	3363	Route 4	672 460	9 0967.59
Cluster 5	37.35	47.29	2988	Route 5	597 451	43 746.91
Cluster 6	21.59	28.94	1800	Route 6	359 931	16 127.55
	200.48	250.32			1 795 748	160 520.62

Table 9. The number of trips and days to support victims (mean speed 40 kph)

Quantity of vehicles	Quantity of supported people	Total trips (Km)	Covered distance by trip	Time (HRS)	Mean correction factor	Corrected time	Days
10	2000	899	225037.7	135022 : 36 : 29	1.76	237639 : 47 : 24	412.56
20	4000	450	112644	67586 : 24 : 00	1.76	118952 : 03 : 50	206.51
50	10 000	180	45057.6	27034 : 33 : 36	1.76	47580 : 49 : 32	82.60
100	20 000	90	22528.8	13517 : 16 : 48	1.76	23790 : 24 : 46	41.30
200	40 000	45	11264.4	6758 : 38 : 24	1.76	11895 : 12 : 23	20.65
600	120 000	18	4505.76	2703 : 27 : 22	1.76	4758 : 04 : 57	8.26

For instance, in Table 9, we obtained the number of vehicles needed to complete the route in an acceptable number of days (8.26 days), using 600 vehicles. It would support 120 000 people with 18 trips (the number of trips is needed in each identified route until complete the requested demand). Finally, as an expected result, we can see in Fig. 2, that the number of supported people will increase with more assigned vehicles.

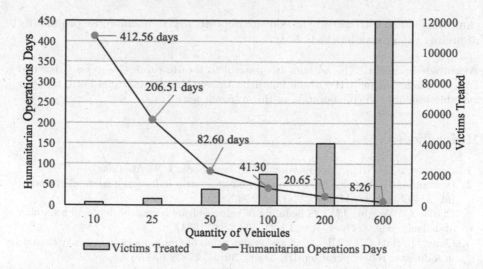

Fig. 2. Number of vehicles vs humanitarian operation days and victims treated

4   Conclusions and Future Works

In this study, we propose an approach to optimize aid distribution kits in a possible disaster in Lima. Previous research works consider the existence of infrastructure, transport, capacity, availability of public services, among others; as a post-disaster state. However, a solution should be suitable to manage an uncertain lack of resources, loss of capacity and infrastructure. Thus, our approach using the corrected distances representing the vulnerability of a location, uses minimal resources (time to complete routes), and it is reliable (routes made under post-disaster conditions). Our results, suggests that the method of Hierarchical Ascendant Classification (HAC), allow us to find an approximate route solution, considering a post-disaster environment.

We found a sufficient and satisfactory humanitarian relief distribution, the searching solution criterion was the shortest time route with the lowest cost, under a spatial configuration which represents a post-disaster state, considering a Correction Factor (CF) to nominal times. This CF was calculated considering a previous HAC analysis, based on vulnerabilities assessment expressed by urbanistic layouts, forecast victims, seismic hazard maps, in Lima and Callao districts. For future research activities, we consider performing an evaluation adding a correction factor which uses resilience assessment and; to evaluate the impact of non-considered costs as manpower, maintenance, resources loading/downloading and security. Furthermore, to carry out a sensitivity analysis to choose particulars trucks, timetables, and outsourcing service.

Also, it is worth mentioning that we will consider the time of loading and discharge each vehicle because this could increase the transportation time. Additionally, a detailed study of different types of vehicles must be follow up, to discover if a different one, could improve the resources usage. Finally, this research can complement with a study to relocate deposits and determine the probability of finding the streets blocked.

Acknowledgement. The authors are grateful to the Research Group on Crisis and Disaster Management (CID) of the Pontificia Universidad Católica del Perú, for their valuable support and collaboration to this research work.

References

1. Assessment, U.N.D., Coordination: Field handbook: UNDAC (2006)
2. Beamon, B.M., Balcik, B.: Performance measurement in humanitarian relief chains. Int. J. Public Sect. Manag. **21**(1), 4–25 (2008)
3. Clarke, G., Wright, J.W.: Scheduling of vehicles from a central depot to a number of delivery points. Oper. Res. **12**(4), 568–581 (1964)
4. Gong, Q., Batta, R.: Allocation and reallocation of ambulances to casualty clusters in a disaster relief operation. IIE Trans. **39**(1), 27–39 (2007)
5. Hall, M.: Supply chain management in the humanitarian context: anatomy of effective relief and development chains. Master thesis, Webster University (2012)
6. Justel, A.: Técnicas de análisis multivariante para agrupación: Métodos cluster (2008)
7. Klibi, W., Martel, A.: Modeling approaches for the design of resilient supply networks under disruptions. Int. J. Prod. Econ. **135**(2), 882–898 (2012)
8. Leseure, M., Hudson-Smith, M., Chandes, J., Paché, G.: Investigating humanitarian logistics issues: from operations management to strategic action. J. Manuf. Technol. Manag. **21**(3), 320–340 (2010)
9. Martinez, A.J.P., Stapleton, O., Van Wassenhove, L.N.: Field vehicle fleet management in humanitarian operations: a case-based approach (2011)
10. Matthew, A., Elliott, R.J., Toshihiro, O., Strobl, E.: Natural disasters, industrial clusters and manufacturing plant survival (2015)
11. Özdamar, L., Demir, O.: A hierarchical clustering and routing procedure for large scale disaster relief logistics planning. Transp. Res. Part E Logist. Transp. Rev. **48**(3), 591–602 (2012)
12. Pidd, M., De Silva, F., Eglese, R.: A simulation model for emergency evacuation. Eur. J. Oper. Res. **90**(3), 413–419 (1996)
13. Prins, C., Lacomme, P., Prodhon, C.: Order-first split-second methods for vehicle routing problems: a review. Transp. Res. Part C Emerg. Technol. **40**, 179–200 (2014)
14. Serpa Oshiro, V.R.: Optimización y localización de almacenes de abastecimiento para la atención de un terremoto de gran magnitud en lima metropolitana y callao (2014)
15. Tai, C.A., Lee, Y.L., Lin, C.Y.: Urban disaster prevention shelter location and evacuation behavior analysis. J. Asian Archit. Build. Eng. **9**(1), 215–220 (2010)
16. Thevenaz, C., Resodihardjo, S.L.: All the best laid plans... conditions impeding proper emergency response. Int. J. Prod. Econ. **126**(1), 7–21 (2010)
17. Toth, P., Vigo, D.: The Vehicle Routing Problem. Monographs on Discrete Mathematics and Applications. Society for Industrial and Applied Mathematics, Philadelphia (2002)
18. Villardón, J.L.V.: Introducción al análisis de clúster. Departamento de Estadística, Universidad de Salamanca. p. 22 (2007)

An Approach to Evaluate Class Assignment Semantic Redundancy on Linked Datasets

Leandro Mendoza[1,2]([⊠]) and Alicia Díaz[2]

[1] CONICET, Consejo Nacional de Investigaciones Científicas y Técnicas,
La Plata, Argentina
[2] LIFIA, Facultad de Informática, Universidad Nacional de La Plata,
La Plata, Argentina
lmendoza@lifia.info.unlp.edu.ar

Abstract. In this work we address the concept of *semantic redundancy* in *linked datasets* considering *class assignment* assertions. We discuss how redundancy can be evaluated as well as the relationship between redundancy and some class hierarchy aspects: number of classes, number of instances a class has, number of class descendants and class depth. Finally, we performed an evaluation on the DBpedia dataset using SPARQL queries for data redundancy checks. Results obtained from this evaluation suggest that the number of redundant class assignments increases when the number of classes is higher, for general classes, with more descendants and for those with more number of instances. In this evaluation we also observed some patterns that can be used to classify class assignments. These observations may be useful for linked data stakeholders to understand how different schemas are used within a dataset, detect errors and improve the mechanisms to generate linked data.

1 Introduction

The amount of interlinked knowledge bases built under *Semantic Web* technologies and following the *linked data* [3] principles has increased significantly last years. These knowledge bases (also known as *linked datasets*) contain information that associates *Web* entities (called *resources*) with well-defined semantics that specifies how these entities should be interpreted. In most linked datasets a substantial amount of data corresponds to *class assignment* assertions, that is, information that specifies resources (or individuals) as instances of certain classes. In this sense, resources are typified using classes usually defined through ontologies and organized into class hierarchies or taxonomies. One of the most important features of *linked datasets* is the ability to link resources between different datasets, which means that resources in one dataset can be described using concepts defined in an external ontology or vocabulary. Besides, several different ontologies can be combined to classify resources within the same dataset giving rise to huge and complex interlinked structures that can suffer from data quality problems [5]. Thus, the use of practical mechanisms to handle knowledge conciseness becomes increasingly important to improve the overall dataset

J.A. Lossio-Ventura and H. Alatrista-Salas (Eds.): SIMBig 2015/2016, CCIS 656, pp. 81–94, 2017.
DOI: 10.1007/978-3-319-55209-5_7

quality and the study of redundancy from a class assignments perspective aims to contribute in this way. From a data quality point of view, redundancy is related with the concept of *extensional conciseness* which has been defined in [14] as "the case when the data does not contain redundant objects at instance level". Thus, considering class assignments, redundancy means that a resource is specified as member of a class when it is not necessary, either because the information is explicitly duplicated or because it can be derived from information that already exists. Current works that have dealt with semantic redundancy on linked datasets implement algorithms based on graph pattern discovering techniques. In contrast, our work proposes a simplified approach based on SPARQL queries and considering class assignment assertions. We discuss how redundancy can be evaluated and perform an evaluation over the DBpedia dataset [8] in order to understand the relationship between redundancy and three class hierarchy aspects: the number of instances a class has, the class depth and its number of descendants. This approach may be useful for linked data users who need to measure semantic data redundancy in a practical way, understand its origin and detect when it may be useful (e.g. to improve performance) or when it can affect negatively the knowledge base (e.g. misuse of classes when typifying resources). The following sections are organized as follows: Sects. 2 and 3 give some background definitions and related work, respectively. Section 4 introduces the redundancy definition adopted and Sect. 5 discusses some mechanisms to address it on linked datasets. Section 6 shows the evaluation results and Sect. 7 describes some patterns to classify class assignments. Finally, some conclusions and further work are given in Sect. 8.

2   Background

In the linked data context, datasets are knowledge bases described using the RDF[1] data model and published following the *linked data principles*[2]. These datasets are collection of assertions about *resources* specified following the "*subject predicate object*" pattern. Assertions are RDF triples and resources may be anything identifiable by an HTTP URI. Knowledge representation mechanisms such as RDFS[3] and OWL[4] extend RDF and allow datasets to be augmented with more expressive semantics. For example, it is possible to describe ontologies by specifying classes and relationships between them (e.g. "*Writer* rdfs:subClassOf *Person*") and to specify resources as member of those classes (e.g. "*William_Shakespeare* rdf:type *Writer*"). From an overall perspective, information contained in these datasets can be split into two levels: *schema level* and *instance level*. *Schema level* refers to *terminological knowledge* (known as TBox), for example, classes, properties and their relationships. On the other hand, *instance level* refers to *assertional knowledge* (known as ABox),

---

[1] https://www.w3.org/RDF/.

[2] https://www.w3.org/DesignIssues/LinkedData.html.

[3] https://www.w3.org/TR/rdf-schema/.

[4] https://www.w3.org/standards/techs/owl#w3c_all.

that is, propositions about entities of a specific domain of interest. An important type of assertional knowledge corresponds to *class assignments*, that is, RDF statements of the form "*resource* rdf:type *class*" used to specify resources as members of certain classes. The most common way to retrieve this information from linked datasets is through a SPARQL[5] endpoint. These endpoints are web services that accept SPARQL queries and return information that match with a given pattern. In this work we will use this mechanism to detect redundant class assignments.

3   Related Work

In the linked data literature, redundancy is related with the data quality dimension of *conciseness* [14] and has been studied and categorized from syntactic to semantic and from schema to instance levels [11]. From a syntactic perspective most of the existing compression techniques focus on RDF serialization. On the other hand, from a semantic perspective, just a few works addressed redundancy. In [13] authors propose a graph based analysis method to identify graph patterns that can be used to remove redundant triples and calculate the volume of semantic redundancy. In [6] authors employ frequent itemset (*frequent pattern*) mining techniques to generate a set of logical rules to compress RDF datasets and then use these rules during decompression. Both works mention the idea of semantic compression by removing derivable knowledge. Regarding the use of SPARQL for quality assessment, [2,7] use query templates to detect some quality problems but semantic redundancy is not included. In [12], authors evaluate instance data in knowledge management systems based on semantic web technologies and use SPARQL queries to identify redundant individual types considering *subclass* relationships. Inspired on the ideas of these works, we use a SPARQL query oriented approach to evaluate redundant class assignments and make this information explicit to users.

4   Redundant Class Assignments

As we know, linked datasets are basically sets of RDF triples and knowledge is specified using mechanisms provided by RDFS and OWL, each one with its own well-defined semantics [4]. In this way, *schema* and *instance* level assertions can be considered as *propositions* to formally describe the notion of derivable knowledge. For example, the notation $\{p_1, p_2\} \models \{p_3, p_4\}$ (where $\models$ is called *entailment relation*) states that propositions p_3 and p_4 (also p_1 and p_2) are logical consequences of propositions p_1 and p_2 obtained under a certain set of rules (logic). Considering this, in previous work [10] we proposed a theoretical framework in which we interpreted the quality dimensions of *redundancy, consistency* and *accuracy* from a semantic perspective and applied it to class assignments.

---

[5] https://www.w3.org/TR/rdf-sparql-query/.

In that work, the concept of *redundancy* is associated to what is known in mathematical logic literature as *independence* [9], that is, the ability to deduce a proposition from other propositions. Formally, given a logic L (semantics) and a set of propositions P, it is defined as *independent* if for all proposition $p_i \in P$ does not hold that $\{P - p_i\} \models p_i$. In this way, a *non-independent* set of propositions can be considered redundant since it contains extra information that if it is removed, it can be obtained from the remaining propositions applying an inference mechanism. Similarly, an *independent set* of propositions can be considered *non-redundant*.

In Sect. 2, we mentioned that *class assignments* are *instance level* assertions (or propositions) that specify resources as members of certain classes. Thus, given a resource r, its class assignment set (CAS_r) contains all the propositions that specify the classes to which r belongs. The idea of the previous paragraph can be applied to *class assignments* to define semantic redundancy since the non-redundant class assignment set of a resource r $(NRCAS_r)$ is the *independent* set of CAS_r. Then, the redundant class assignment set $(RCAS_r)$ can be considered as the difference of those sets. For example, consider the following *schema level* propositions about classes *Agent*, *Person* and *Writer*:

- p_1: *Writer* rdfs:subClassOf *Person*,
- p_2: *Person* rdfs:subClassOf *Agent*,

Then, consider the following CAS_r for a resource r:

- p_3: r rdf:type *Agent*
- p_4: r rdf:type *Person*
- p_5: r rdf:type *Writer*

As we can see, the given CAS_r is redundant since if we remove propositions p_3 and p_4 from CAS_r, they can be deduced from the remaining proposition p_5 by transitivity of rdfs:subClassOf relationship. On the other hand, if p_5 is removed, it can not be obtained from propositions p_3 and p_4. Thus, the *non-redundant class assignment set* for the resource r $(NRCAS_r)$ would be $\{p_5\}$ and the $RCAS_r$ would be $\{p_3, p_4\}$.

5   Mechanisms to Check Class Assignment Redundancy

Given a *linked dataset*, we propose a four step process to evaluate class assignment redundancy:

- **Step 1**. Identify assertional knowledge that corresponds to class assignments. This information refers to assertions that follow the pattern *"resource,* rdf:type, *class"*.
- **Step 2**. Identify referenced ontologies. Determine to which ontology the classes detected in the previous step belong.
- **Step 3**. Identify class assignments sets by grouping assignments per resource. Those class assignments that have the same resource (r) as subject, belong to the same group and conform the class assignment set of r (CAS_r).

– **Step 4**. Evaluate redundant class assignments considering the resource r that describe and the class assignment set (CAS_r) to which they belong. Then, each class assignment can be classified into two groups: as redundant $(RCAS_r)$ or not $(NRCAS_r)$.

In the following subsections, we discuss some techniques that can be used to compute $NRCAS_r$ (and $RCAS_r$) on linked datasets. Then, in Sect. 6, we will implement the evaluation process and will use one of these techniques to perform our evaluation.

5.1 Redundancy Checks Using SPARQL

Using SPARQL, a simple query can be implemented to get the $NRCAS$. For example, query in listing 1 can be used to get the non-redundant set of classes of a given resource (specified by *resource_URI*).

```
SELECT DISTINCT ?c
WHERE {
 <resource_URI> rdf:type ?c
 FILTER regex(str(?c),"ont_URI","i")
 FILTER NOT EXISTS {
   <resource_URI> rdf:type ?sc .
   FILTER regex(str(?sc),"ont_URI","i")
   ?sc rdfs:subClassOf ?c }
}
```

Listing 1: SPARQL query example to get non-redundant class assignments

Note that the mentioned query example considers only one ontology (filtered by *ontology_URI*) and does not implement any inference mechanism at instance or schema level. This means that the query will work while all class assignments and relationships between the involved classes will be specified explicitly on the dataset. If this is not the case and a transitive closure of sub/super classes is needed, it is necessary to implement an algorithm that iterates recursively over these queries until it gets the required classes. Using SPARQL *property paths*[6] (e.g. `rdfs:subClassOf*` or `rdfs:subClassOf+`) it is possible to check connectivity of two classes by an arbitrary length path (route through a graph between two graph nodes). For example, we can get all the classes that are descendant (or subclasses) of a given class specified by a URI (<class_URI>) by counting classes that match the pattern "`rdf:subClassOf*` <class_URI>". In a similar way, we can get all the ancestors of a given class by counting classes that match the pattern "<class_URI> `rdf:subClassOf*` ?c". It is important

---

[6] https://www.w3.org/TR/sparql11-property-paths/.

to highlight that the performance of SPARQL queries depends on its implementation and the dataset size. Although complex SPARQL queries can become unacceptably slow when working with large amounts of data, it is currently the most practical mechanism to access linked datasets.

5.2 Redundancy Checks Using Graph Based Algorithms

Given a class hierarchy, a resource r and its CAS_r, a way to compute redundancy is by interpreting the class hierarchy as a directed acyclic graph "G" (or a tree) in which each node is a class and each edge is the relation rdfs:subClassOf. A node "A" of "G" can be considered a class if there exist a triple with the form "A rdfs:subClassOf x", "x rdfs:subClassOf A" or "x rdf:type A". Then, a class "B" is subclass of a class "A" if node "A" is reachable from node "B" in "G", that is, if exists a path between B and A in the graph. Considering this, given a proposition set Q that specifies the classes to which an instance i belongs, a naive algorithm can be implemented to compute a *non-redundant* proposition set R: first set $R = Q$, then for each element q in Q check if there is a path from some of the remaining propositions in Q to q, if so, q is deleted from R. Finally, the algorithm returns R which is then the *non redundant class assignments set* of r. Removing redundancies can be associated with the problem known as *transitive reduction* [1] which has an unfortunate complexity class if it is implemented naively. The main drawbacks of these kind of algorithms are the lack of flexibility to deal with heterogeneous and evolving data and the impossibility to exploit the underlying semantics.

5.3 Redundancy Checks Using Reasoners

Another alternative to compute redundancy is by using *Semantic Web* reasoners based on decidable fragments of RDFS and OWL semantics. These tools implement inference mechanisms that can be used to deduce if a resource belongs to a given class (instance checking). The main advantage of using reasoners is that the potential of the underling semantics can be exploited (e.g. several ontologies can be combined to get implicit knowledge). On the other hand, the disadvantage of using these tools in the linked data scenario is its complexity: inference techniques work well for small examples with limited knowledge but they turn unacceptably slow for large-scale datasets. Besides, when multiples ontologies are combined, inconsistencies can arise affecting the inference process and hampering the detection of redundant propositions.

6 Evaluation

To perform our evaluation we selected the English version of DBpedia[7] and set up a local mirror using a Virtuoso[8] server (version 7.2). The mechanisms implemented to compute redundant class assignment avoid the use of complex graph

---

[7] http://wiki.dbpedia.org/Downloads2015-04.
[8] http://virtuoso.openlinksw.com/.

based algorithms or RDFS/OWL reasoners and use a SPARQL query oriented approach (see Sect. 5.1). We follow a class assignment oriented evaluation process (see Sect. 5): (i) we got resources that belong to classes of specific schemas or ontologies, (ii) for each resource we got its class assignment set and (iii), for each class assignment contained in those sets we analyzed which of them were redundant with respect to its referenced schema. Finally, for those redundant class assignments, we analyzed some characteristics of the classes they refer .

Although resources in DBpedia are classified using several classes of different schemas we only considered the DBpedia[9] core and YAGO[10] ontologies because information about the involved class hierarchies (subclass relationships) can be obtained directly from queries through the dataset SPARQL endpoint. DBpedia ontology is a shallow cross-domain ontology that covers more than 600 classes and was created based on Wikipedia *infoboxes*. YAGO is a taxonomy used in the YAGO knowledge base that currently covers more than 350,000 classes. The evaluation is organized in the next subsections as follows: we first performed an overall redundancy evaluation considering DBpedia and YAGO ontologies and then a further analysis was done per class groups but only considering the DBpedia class hierarchy in order to keep the number of classes manageable. Classes were grouped considering three characteristics: depth in the class hierarchy, number of class descendants and number of class assignments (or instances). For each case, we analyzed its relationship with redundancy.

6.1   Overall Redundancy Evaluation

The first overall evaluation was made by retrieving all resources that belong to some class of the DBpedia ontology (6,729,604 resources of 453 classes) and then we did the same with the YAGO ontology (2,886,306 resources of 369,144 classes). For each resource we compute its CAS, its $NRCAS$ and its $RCAS$ (see Sect. 4) considering both ontologies separately. Information about resources and its CAS and $NRCAS$ were obtained through SPARQL queries (see Sect. 5.1) and $RCAS$ was obtained by computing the difference $CAS - NRCAS$. Results can be viewed in table of Fig. 1. Each element of each CAS was counted as a different class assignment ($nbCA$ column), each element of $NRCA$ was counted as a non-redundant class assignment ($nbNRCA$ column) and each element of RCA was counted as a redundant class assignment ($nbRCA$ column). As we can see in chart of Fig. 1, considering classes of the DBpedia ontology almost half class assignments are redundant. On the other hand, considering the YAGO ontology 80% of class assignments are redundant. In the latter case, the amount of class assignments is higher and the amount of concepts in the class hierarchy increases considerably. These results suggest a relationship between the number of classes, class assignments and redundancy: as the number of classes and class assignments increases, so the probability of redundancy.

---

[9] http://wiki.dbpedia.org/services-resources/ontology.

[10] https://www.mpi-inf.mpg.de/departments/databases-and-information-systems/ research/yago-naga/yago/.

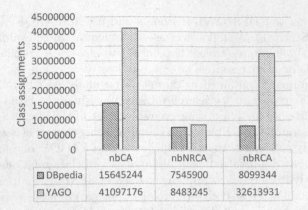

Fig. 1. DBpedia and YAGO redundancy evaluations

6.2 Redundancy and Class Depth

To analyze the relationship between redundant class assignments and class depth we categorized classes into groups from 0 to 6 according to their depth in the DBpedia class hierarchy (the distance from the root to that class) and then we count how many class assignments refer to those classes. Classes with depth 0 are the most general and 6 is the max depth found in the DBpedia class hierarchy. A class assignment refers to (or belongs to) a class C if it is a triple of the form ($resource$ rdf:type C). To compute the depth of a class we used a SPARQL query to count the number of ancestors (see Sect. 5.1). Results can be viewed in Table 1 and chart of Fig. 2 shows the relationship between the class depth and the percentage of redundant class assignments (% of RCA).

Table 1. Redundancy and class depth evaluation.

Depth	nbClasses	nbCA	nbRCA
0	32	5,037,966	3,940,920
1	86	3,941,230	2,877,434
2	110	5,385,831	1,156,327
3	180	1,154,660	118,886
4	39	118,891	5,777
5	5	5,777	0
6	1	889	0

From Table 1, we can observe how redundant class assignments are distributed: almost 98% of the total number of redundant class assignments refer to more general classes (with depth 0, 1 and 2), which are half of the classes considered (228). In chart of Fig. 2 we can see that as the class depth increases

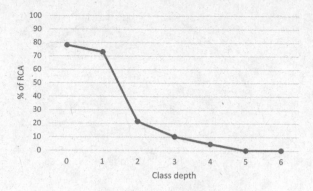

Fig. 2. RCA vs class depth

(more specific a class is), the relative number of redundant class assignments that refers to those classes decreases. For example, we can see that there are 32 classes (*nbClasses* column) of depth 0 (*Depth* column), 5,037,966 class assignments (*nbCA* column) that refer to those classes and 3,940,920 of them (78%) are redundant (*nbRCA* column). Examples of classes that belong to that group are *Agent*, *Place*, *Work*, etc.

6.3 Redundancy and Class Descendants

To analyze the relationship between redundant class assignments and class descendants we categorized classes into 10 groups according to the number of descendants that they have. To compute the descendants we used a SPARQL query to get the subclasses of a given class (see Sect. 5.1). Results are showed in Table 2 and chart of Fig. 3 shows the relationship between the number of class

Table 2. Redundancy and class descendants evaluation.

nbDesc	nbClasses	nbCA	nbRCA
0	330	3, 250, 543	0
1 to 5	78	3, 209, 728	321, 078
6 to 10	16	948, 187	524, 262
10 to 20	13	666, 630	531, 848
21 to 30	8	773, 316	751, 488
31 to 50	2	535, 606	502, 892
51 to 70	3	809, 171	784, 103
71 to 100	1	220, 219	211, 415
101 to 200	2	2, 860, 585	2, 117, 098
More than 200	1	5, 231, 844	4, 472, 258

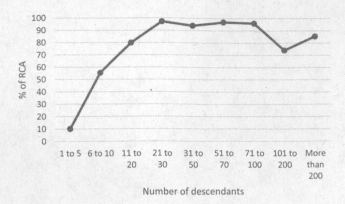

Fig. 3. RCA vs class descendants

descendants and the percentage of redundant class assignments (% of RCA). Classes that do not have descendants (330 classes) are the most specific and class assignments that belong to that group are not redundant. As we can see in chart of Fig. 3, when the number of descendant per class increases, the number of redundant class assignments that refer to those classes also increases. For example, group named "1 to 5" refers to classes that have between 1 to 5 descendants (78 classes) and redundancy is relatively low. On the other hand, assignments that refer to classes with several descendants (e.g. class *Agent*) have a high level of redundancy. In Table 2 we can also observe that there are only 3 classes with more than 100 descendants (*Agent*, *Person* and *Place*) and they concentrates almost the 65% of redundant class assignments.

6.4 Redundancy and Number of Class Assignments

To analyze the relationship between redundant class assignments and the number of class assignments per class we categorized classes into 10 groups according to the number of class assignments that refers to a class. To do this, we counted the number of different resources (instances) that belongs to some specific class using a SPARQL query. The aim of this analysis was to observe if those classes with more instances were also the classes with more redundant class assignments. Table 3 shows the evaluation results and chart of Fig. 4 shows the relationship between redundancy and the number of class assignments (or instances) per class. Columns *nbCA-acc* and *nbRCA-acc* show the total number of class assignments and redundant class assignments in each group.

As we can see in Table 3, most of classes (315) have less than 10 K class assignments. Besides, as this number increases the number of classes involved decreases but the relative percentage of redundant class assignments increases (see chart in Fig. 4). For example, classes that have more than 500 K class assignments (e.g. *Agent*, *Place*, *Person*, etc.) concentrate most of them (9,292,013) and 48% are redundant. We also observe that the second group of most used classes

Table 3. Redundancy and number of class assignments evaluation.

nbCA	nbClasses	nbCA-acc	nbRCA-acc
0 to 10 K	315	599, 261	101, 861
10 K to 20 K	35	484, 931	120, 512
20 K to 30 K	34	588, 845	267, 460
30 K to 40 K	25	384, 601	220, 270
40 K to 30 K	11	900, 290	36, 156
100 K to 200 K	7	906, 730	365, 085
200 K to 300 K	5	1, 274, 010	1, 222, 844
300 K to 400 K	1	396, 046	395, 804
400 K to 500 K	2	934, 843	681, 445
More than 500 K	6	9, 292, 013	4, 472, 258

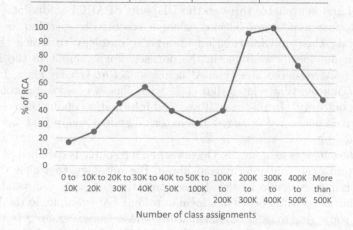

Fig. 4. RCA vs number of class assignments

(between 200 K and 500 K class assignments) consists of about 8 classes with high levels of redundancy (between 70% and 95%).

7   Class Assignments Classification

Considering the evaluation results, we observed some common situations that can be used to classify class assignments according to two main aspects: the relationship between classes (subclass and equivalence) of a class assignment set and if the information about those classes is available locally or not. Considering the first aspect, we distinguished two assignment patterns:

– *Super-class assignment.* Occurs when a resource is instance of a certain class and is also specified as member of some super-classes of that class. For example, in the DBpedia dataset, class *dbo:Writer* is subclass of *dbo:Person* and a resource is specified as member of both classes.

– *Equivalent class assignment.* Occurs when a resource is instance of a certain class and is also specified as member of other classes defined as equivalent to that class. For example, in the DBpedia dataset, class *dbo:Writer* is defined as equivalent to *wikidata:Q36180*[11] and a resource is specified as member of both classes.

Depending on whether the resource is typified using classes from a local ontology (schema information is available) or is described using classes of external ontologies (defined in other linked dataset), we distinguished three assignment patterns:

– *Local class assignment.* Occurs when a resource is specified as instance of a certain class and the information about the ontology to which that class belongs is available in the same dataset. For example, in the DBpedia dataset, information about classes defined in the DBpedia and YAGO ontologies was imported and is available through the DBpedia SPARQL endpoint.
– *External class assignment.* Occurs when a resource is specified as instance of a certain class and information about the ontology to which that class belongs is not locally available in the dataset. For example, in the DBpedia dataset, some resources are typified using *umbel-rc:Artist*[12] class, but there is no information that relates that class with classes of a local ontology (e.g. DBpedia ontology). In this case, if we need information about that class or its schema, it is necessary to access the external source where that information is defined.
– *Semi-external class assignment.* Occurs when a resource is specified as instance of a certain class and information about the ontology to which that class belongs is not available in the dataset but there is some information that relates that class with classes of a local ontology. For example, in the DBpedia dataset, some resources are typified using the *schema:Person*[13] but there is no schema information about it in DBpedia. Despite this, *schema:Person* is defined as equivalent to *dbo:Person* class (defined in a local ontology) and this information allows the inference of some schema knowledge about the external class.

The above classification may be useful to understand which is the origin of semantic redundancy (e.g. general or equivalent assignments) and to discover isolated assertions (e.g. meaningless assignments due to unreachable schema information).

8   Conclusions and Future Work

In this work we addressed the concept of *semantic redundancy* considering *class assignments* assertions in *linked datasets*. Based on a formal definition

---

[11] https://www.wikidata.org/wiki/Q36180.
[12] UMBEL reference concepts http://umbel.org/.
[13] http://schema.org/Person.

we discussed how *redundant* (and *non-redundant*) class assignments sets can be detected. Inspired in previous related work, we conducted an evaluation over the English version of DBpedia based on SPARQL queries. We analyzed the relationship between redundancy and some class hierarchy characteristics: the number of classes, the number of instances a class has, its depth and its number of descendants.

Regarding the evaluation results, they suggest that there is a relationship between redundancy and depth of a class in a hierarchy: as more general a class is (less depth), more redundant class assignments can be found that refer to that class. In a similar way, as the number of descendants per class increases so does the number of redundant class assignments related to that class. Particularly, we noted that this also occurs when the number of classes is higher and also for most populated classes (with more class assignments). In this sense, datasets that use complex and large class hierarchies to typify their resources may be more prone to *class assignment* redundancy. Considering this, we make the following additional observations:

- Redundancy analysis can be used to detect class assignments patterns and data publishers behaviors. For example, in our evaluation we detected that when a resource is assigned to a very specific class, it is also assigned explicitly to the ancestors of that class. Discovering these kinds of patterns may be useful to improve the linked data generation process or even to understand how classes described in a given ontology are used on a specific dataset.
- SPARQL queries as a mechanism to evaluate redundancy offer practical ways to implement quality checks and get some statistics of linked datasets. However, inference capabilities are limited: we can discover just some graph patterns using queries but other implicit knowledge will be *unreachable* since we can not exploit all the semantic capabilities or expressiveness of the used languages.

Future work will be focused on the development and assessment of semantic redundancy metrics that support our results on other linked datasets. Besides, we plan to study how redundancy can affect other data quality dimensions particularly those related with semantic accuracy.

References

1. Aho, A.V., Garey, M.R., Ullman, J.D.: The transitive reduction of a directed graph. SIAM J. Comput. **1**(2), 131 (1972)
2. Fürber, C., Hepp, M.: Using semantic web resources for data quality management. In: Cimiano, P., Pinto, H.S. (eds.) EKAW 2010. LNCS (LNAI), vol. 6317, pp. 211–225. Springer, Heidelberg (2010). doi:10.1007/978-3-642-16438-5_15
3. Heath, T., Bizer, C.: Linked Data: Evolving the Web into a Global Data Space, Synthesis Lectures on the Semantic Web: Theory and Technology, vol. 1, 1st edn. Morgan Claypool, Palo Alto (2011). html version edition
4. Hitzler, P., Krötzsch, M., Rudolph, S.: Foundations of Semantic Web Technologies. Chapman & Hall/CRC, Boca Raton (2009)

5. Hogan, A., Harth, A., Passant, A., Decker, S., Polleres, A.: Weaving the pedantic web. In: Linked Data on the Web Workshop (LDOW 2010) at WWW 2010, vol. 628, pp. 30–34. CEUR Workshop Proceedings (2010)
6. Joshi, A.K., Hitzler, P., Dong, G.: Logical linked data compression. In: Cimiano, P., Corcho, O., Presutti, V., Hollink, L., Rudolph, S. (eds.) ESWC 2013. LNCS, vol. 7882, pp. 170–184. Springer, Heidelberg (2013). doi:10.1007/978-3-642-38288-8_12
7. Kontokostas, D., Westphal, P., Auer, S., Hellmann, S., Lehmann, J., Cornelissen, R., Zaveri, A.: Test-driven evaluation of linked data quality. In: Proceedings of the 23rd International Conference on World Wide Web (WWW 2014), pp. 747–758. International World Wide Web Conferences Steering Committee, Republic and Canton of Geneva (2014)
8. Lehmann, J., Isele, R., Jakob, M., Jentzsch, A., Kontokostas, D., Mendes, P.N., Hellmann, S., Morsey, M., van Kleef, P., Auer, S., Bizer, C.: DBpedia - a large-scale, multilingual knowledge base extracted from Wikipedia. Semant. Web J. 6(2), 167–195 (2015)
9. Mendelson, E.: Introduction to Mathematical Logic, 5th edn. Chapman & Hall/CRC, Boca Raton (2009)
10. Mendoza, L., Díaz, A.: Adequate class assignment on linked data. In: Proceedings of the 2016 IEEE/WIC/ACM International Conference on Web Intelligence. IEEE Computer Society (2016)
11. Pan, J.Z., Gómez-Pérez, J., Ren, Y., Wu, H., Zhu, M.: SSP: compressing RDF data by summarisation, serialisation and predictive encoding. Technical report, 07 2014 (2014). http://www.kdrive-project.eu/resources
12. Tao, J., Ding, L., McGuinness, D.L.: Instance data evaluation for semantic web-based knowledge management systems. In: Proceedings of the 42nd Hawaii International Conference on System Sciences (HICSS 2009), pp. 1–10 (2009)
13. Wu, H., Villazn-Terrazas, B., Pan, J.Z., Gómez-Pérez, J.M.: How redundant is it? - An empirical analysis on linked datasets. In: Hartig, O., Hogan, A., Sequeda, J. (eds.) COLD, vol. 1264. CEUR Workshop Proceedings. CEUR-WS.org (2014)
14. Zaveri, A., Rula, A., Maurino, A., Pietrobon, R., Lehmann, J., Auer, S.: Quality assessment for linked data: a survey. Semant. Web 7(1), 63–93 (2016)

Topic-Based Sentiment Analysis

Prasadith Buddhitha$^{(\boxtimes)}$ and Diana Inkpen

School of Electrical Engineering and Computer Science,
University of Ottawa, Ottawa, ON K1N6N5, Canada
{pkiri056,Diana.Inkpen}@uottawa.ca

Abstract. We present a method that exploits syntactic dependencies for topic-oriented sentiment analysis in tweets. The proposed solution is based on supervised text classification and available polarity lexicons in order to identify the relevant dependencies in each sentence by detecting the correct attachment points for the polarity words. Our experiments are based on the data from the Semantic Evaluation Exercise 2015 (SemEval-2015), task 10, subtask C. The dependency parser that we used is adapted to this kind of text. Our classifier that combines topic- and sentence-level features obtained very good results.

1 Introduction

The growth of social media has enhanced the amount of information being shared among people. Opinions about various topics are among the most important pieces of information exchanged. Microblogging platforms are used due to their accessibility and short messages. Twitter has become one of the most popular microblogging platforms in recent years. As more and more interest has emerged in identifying the key information contained within the messages, greater difficulties are being introduced due to the informal nature of the message representation. With the limitation of 140 characters, the informal nature of the messages has introduced slang, new words, URLs, creative spelling, misspelling, punctuation's and abbreviations such as #hashtag and "re-tweet" (RT).

With the representation of valuable information about one or more interests enriched with user perception and the sheer amount of volume has challenged the researchers in Natural Language Processing and Machine Learning to generate mechanisms to extract valuable information, which could be beneficial for the interested parties from different domains, such as marketing, finance, and politics. Identification of the perception, which could also be termed as opinion mining or sentiment analysis has resulted in many types of research based on supervised and unsupervised learning methods.

The widely-spread enthusiasm in the Twitter sentiment analysis is supported by various research-based events such as the Semantic Evaluation Exercise (SemEval). The research we present in this paper is based on the SemEval 2015 task 10, dedicated to Sentiment Analysis in Twitter. The task is subdivided into four sub-tasks emphasizing different levels such as expression, message, topic and trend [24].

© Springer International Publishing AG 2017
J.A. Lossio-Ventura and H. Alatrista-Salas (Eds.): SIMBig 2015/2016, CCIS 656, pp. 95–107, 2017.
DOI: 10.1007/978-3-319-55209-5_8

We focus on "topic-based message polarity classification"; that is, given a message and a topic, we classify whether the message is of positive, negative, or neutral sentiment towards the given topic [24]. The task will be approached through the use of sentiment lexicons at both topic and sentence level. Our solution will use several freely available general-purpose sentiment lexicons and tweet-specific sentiment lexicons, the latter provided by the National Research Council (NRC) of Canada.

The following paragraphs will briefly define the different terminologies being used in rest of this paper.

Tokenization: Text normalization is a key part in Natural Language Processing (NLP). Tokenization can be considered as one of the initial and key functions in text normalization where a given text is divided into a sequence of characters, which can be treated as a group. The character sequence can be treated as a word or any other token such as a punctuation mark or a number [8].

Sentiment analysis: As described in Scherer's typology of affective states, sentiment analysis can be defined as detecting attitudes [23]. The polarity can be identified as a type of attitude, which could be categorized into one of the states such as positive, negative or neutral, as well as being assigned a weighted value indicating the strength within the assigned category [12].

N-grams: N-grams can be broadly defined as a contiguous sequence of words [8]. The N-grams can be represented as N-tokens, where the tokens could be words, letters, etc. Depending on the number of tokens, N-gram models can be termed as unigrams (N-gram with the size of one), 2-gram (bigram), 3-gram (trigram), four-gram or five-gram, which can be considered as the most commonly used in statistical language models. The number of items within the language model can be based on the processing task. Our proposed solution mainly considers unigrams and bigrams.

Decision Trees: Decision trees can be explained in the most abstract form of if-then-else statements arranged in a tree. The most informative features extracted from the training data are according to their information gain [21]. They have the advantage that the model learned is interpretable; the user can inspect the tree in order to understand why a certain decision was made. Decision trees do not always get the best results compared to other machine learning algorithms (but they happened to work very well for our particular task). The key step in making decision trees effective will be the selection of suitable features for our task. In our solution, the selected features are based on the polarity words from the sentence that are in a dependency relation with the targeted topic.

2   Related Work

There is a lot of research on sentiment analysis. It started with identifying subjective and objective statements. The subjective statement can further be classified into positive, negative, or neutral, possibly with intensity levels for the first two.

Many researches have been done on opinion mining and sentiment analysis for customer reviews [18] and, more recently, on Twitter messages [2,7,11,17].

Many techniques have been adopted by researchers on Twitter sentiment analysis, such as lexical approaches and machine learning approaches [5,27]. Lexicon-based systems focused on creating repositories of words labeled with polarity values, possibly calculated based on the association of these words and with other words with known polarities [5]. In addition, well-performing hybrid systems have also been proposed for Twitter sentiment analysis by combining hierarchical models based on lexicons and language modeling approaches [6].

The large impact of using polarity lexicons in supervised learning can also be seen in the top seven-ranked participants in the SemEval-2015, task 10, sub-task C. According to [3,20,25,28] put emphasis on publicly available lexicons such as the NRC Hashtag Sentiment Lexicon, the Sentiment 140 Lexicon, the NRC Emotion Lexicon and SentiWordNet for feature engineering. In addition to lexicon features, many of the top scored systems used linguistic and Twitter-specific features. These systems have mainly used supervised machine learning implemented through classifiers such as Support Vector Machine (SVM) and logistic regression to obtain the best results. It is interesting to note that [25], ranked sixth for subtask C, have used the Stanford parser configured for short documents with the use of a caseless parsing model. The authors have argued that TweeboParser [10], which is explicitly created for parsing Twitter messages, lacks in dependency type information due to the use of a simpler annotation scheme rather than using an annotation scheme like Penn Treebank. [10] have argued that Penn Treebank annotations produce low accuracies specifically with informal text such as tweets and it is more suited for structured data, and due to this reason they have used a syntactically-annotated corpus of tweets (TWEE-BANK). Despite these claims, the TweeboParser has achieved an accuracy over 80% on unlabelled attachments. The parser has contributed nearly 10% accuracy increase in our proposed solution through topic-level feature extraction.

As many effective sentiment analysis solutions are based on machine learning and lexicon-based techniques [6], our proposed solution will also be focused on supervised machine learning that uses features computed by using freely available lexicons, while focusing on general and Twitter-specific language constructs.

Many of the proposed solutions in sentiment analysis have used key natural language processing techniques such as tokenizing, part-of-speech tagging, and bag-of-words representation for preliminary preparation tasks [9,15,27]. Due to the informal nature of the Twitter messages, text-preprocessing techniques have to be given special consideration. [9,15,27] used the Carnegie Mellon University (CMU) ARK tools for tasks such as tokenizing and part-of-speech tagging, which handles text with Twitter-specific characteristics such as identifying special characters and tokens according to Twitter-specific requirements [16]. In addition to the CMU ARK tokenizer, our proposed solution uses the TweeboParser for Twitter text dependency parsing, which allows us to identify the syntactic structure of the Twitter messages.

It could be argued that supervised or semi-supervised machine learning techniques provide higher accuracy levels compared to unsupervised machine learning techniques and also the consideration must be given to the specific domain which the task is implemented on [26]. This is why we build a supervised classifier based on the SemEval training data, and we are planning to extend it in future work with a large amount of unlabeled Twitter data.

Little consideration was given to hashtags, but this has changed recently, as their impact on the sentiment value of a message was demonstrated. Research has been conducted by using hashtags as seeds representing positive and negative sentiment [11] and also by creating hashtag sentiment lexicons from a hashtag-based emotion corpus [13]. The same lexicon created by [13] is being used by our proposed classifier to identify hashtags associated with opinions and emotions; we add a stronger emphasis on the hashtag representation.

According to [29], emoticons are also considered to be providing considerable sentiment value towards the overall sentiment of a sentence. Emoticons were identified using different mechanisms such as through the use of Christopher Potts' tokenizing script [14]. Our proposed solution has adopted the MaxDiff Twitter sentiment lexicon to identify both the emoticons and their associated sentiment values [9], as it will be described in Sect. 4.2.

Many proposed solutions normalize the informal messages in order to assist in sentiment identification [27, 29]. We do not need to do this, because the lexicons we used contain many such Twitter-specific terms and their associated sentiment values [9, 14].

3   Data

The dataset is obtained from the SemEval-2015 Task 10 for subtask C[1]. The dataset consist of trial and training data. The training data includes the Twitter ID, the target topic and the polarity towards it [24]. Due to privacy reasons, the relevant Twitter messages were not included and a separate script has being provided in order to download the messages. After executing the script, the message "Not Available" is being printed if the relevant tweet is no longer available.

Our final dataset contains 391 Twitter messages, out from 489 given Twitter IDs for the task, where 96 IDs were removed due to unavailability of the Twitter messages, one record due to a mismatched ID and one record because it was a duplicate ID. The original dataset included around 44 topics and approximately ten tweets per topic [24]. From the extracted 391 tweets, 110 tweets were labeled with positive topic polarity, 44 as negative, 235 as neutral and 2 were off-topic. According to [22], having access to less training tweets does not have a substantial impact on the results being generated, because the task participants who used less training data have produced higher results.

In order to make the dataset more relevant and accurate, both URLs and usernames were normalized, where the URLs are renamed as http://someurl and

---

[1] We did not participate in the task, we downloaded the data after the evaluation campaigned.

the usernames as @someuser. The tweets were also tokenized using the tokenizing tool provided by Carnegie Mellon University (CMU), known as Tweet NLP.

The Twitter messages in our dataset were composed of only one sentence (and one target topic in the message), this is why in this paper, the terms "sentence-level" and "message-level" are used interchangeably. This is due to the short nature of the tweets (also they are rarely fully grammatical sentences due to the informal communication style). In a rare case, when a tweet might contain more than one sentence, for future test data, our method will use only the sentence(s) that contain the topic(s) of interest.

4   Experiments

Our experiments had the goal of building a supervised classifier that can decide whether the message is positive, negative or neutral towards the given topic. The experiments were conducted in two parts where features were extracted at the sentence level and topic level, using different lexicons. The following sections will describe our features and the tools that we used to extract them, mainly the dependency parser and the lexicons.

4.1   Dependency Parsing

The dependency parser for English tweets, TweeboParser from CMU, was used to generate the syntactic structure of each tweet. Given an input, one tweet per line, an output of the tokenized tweet is generated with their associated part-of-speech tags and syntactic dependencies. The generated prediction file is structured according to the "CoNLL" format representing different columns such as, token position (ID), word form (FORM), coarse grained part-of-speech tag (CPOSTAG), fine grained part-of-speech tag (POSTAG), most importantly the head of the current token (HEAD) indicating the dependencies and the type of dependency relation (DEPREL) [4]. The generated syntactic structure for the following tweet: "They say you are what you eat, but it's Friday and I don't care! #TGIF (@ Ogalo Crows Nest) http://t.co/l3uLuKGk" is presented in Table 1. For this example, there are several conjunctions (CONJ), and one multiword expression (MWE) is identified. Some other dependency relations were missed in this case, due to the imperfect training of the parser on small amounts of Twitter data.

This example tweet is from our dataset, and according to the annotations provided by the SemEval task, the target topic is "Crows Nest", the general message polarity is "positive", and the polarity towards the topic is "neutral".

4.2   Feature Extraction

Feature extraction was conducted at the sentence level and at the topic level. Feature extraction was mainly based on sentiment lexicons. NRC Canada has provided several tweet-specific sentiment lexicons, which were used in capturing

Table 1. TweeboParser output for a tweet.

ID	FORM	POSTAG	POSTAG	HEAD	DEPREL
1	They	O	O	2	-
2	Say	V	V	9	CONJ
3	You	O	O	4	-
4	Are	V	V	2	-
5	What	O	O	7	-
6	You	O	O	7	-
7	Eat	V	V	4	-
8	,	,	,	−1	-
9	But	&	&	0	-
10	It's	L	L	9	CONJ
11	Friday	∧	∧	10	-
12	And	&	&	0	-
13	I	O	O	14	-
14	Don't	V	V	12	CONJ
15	Care	V	V	14	-
16	!	,	,	−1	-
17	#TGIF	#	#	−1	-
18	(@	P	P	0	-
19	Ogalo	∧	∧	21	MWE
20	Crows	∧	∧	21	MWE
21	Nest	∧	∧	18	-
22	)	,	,	−1	-
23	http://t.co/l3uLuKGk	U	U	−1	-

Twitter specific content displaying different levels of polarity such as positive, negative and neutral, and also accompanied by a finite set of values representing the evaluative intensity towards specific polarity categories [9]. Mentioned below are the different lexicons being used at both sentence and topic levels.

The *NRC hashtag emotion lexicon* consists of a list of words and their association with eight emotions: anger, fear, anticipation, trust, surprise, sadness, joy and disgust. The association between the tweets and the emotions were calculated through the identification of emotion-word hashtags in tweets [14]. The file is formatted according to category (e.g.m anger, fear, etc.), the target word, and the associated score. The relevant score indicates the strength of the association between the category and the target word [14]. Higher scores indicate stronger associations between the category and the target word [13].

The *NRC word-emotion association lexicon* contains a list of words and their association with eight emotions, anger, fear, anticipation, trust, surprise, sadness, joy and disgust, and also the polarity towards the relevant word represented either as positive or negative [14]. The lexicon is structured according to the target word, the emotion category and the association value indicating to which category the word belongs. The value 1 indicates that it belongs to the relevant category; the value is 0 if it does not [14].

The *MaxDiff Twitter sentiment lexicon* represents unigrams with associative strength towards positive sentiment. The data was obtained by manual annotation through Amazon Mechanical Turk [9]. Each entry of the lexicon consists of the term and its relevant associative values ranging from −1 indicating the most negative score and +1 indicating the most positive score [14].

Sentiment140 lexicon is a collection of words with the associated positive and negative sentiment [14]. The lexicon is divided into unigrams and bigrams, where each entry contains the term, the sentiment score and the number of times the term appeared with a positive emoticon and the number of times the term appeared with a negative emoticon. The sentiment score is calculated using the pointwise mutual information (PMI), by subtracting the score of the term with negative emoticons from the score of the terms with positive emoticons [14].

SentiWordNet 3.0 was designed for assisting in opinion mining and sentiment classification in general (not for Twitter messages). SentiWordNet is a result of annotating WordNet synonym entries according to their polarity weighting [1]. The scores given for positive, negative and neutral classes range between zero and one, and the summation of all three scores is 1. SentiWordNet 3.0 is based on WordNet version 3.0 and the entries include POS and ID columns identifying the associated WordNet synonym set.

Similar to the "MaxDiff Twitter sentiment lexicon", we used another lexicon with an associative strength ranging from 0 (most negative) to 1 (most positive) in compared to the one already being used, which has an associative score ranging from −1 (most negative) to +1 (most positive).

NRC Hashtag sentiment lexicon contains a list of words with their associated sentiment scores. The score is either a positive or a negative number reflecting the level of association with either positive or negative sentiment. Both bigram and unigram lexicons were used as separate features where the positive and negative sentiment scores were calculated by subtracting the pointwise mutual information (PMI) value of the term associated with negative hashtags from positive hashtags [14].

Sentiment140 Affirmative Context lexicon and *Sentiment140 Negated Context lexicon* consists of both unigram and bigram terms where the sentiment score is calculated by subtracting the PMI value of a term associated with a positive sentiment from a negative sentiment [9].

NRC Hashtag Affirmative Context Sentiment lexicon and *NRC Hashtag Negated Context Sentiment lexicon* were used in both unigram and bigram forms, where the sentiment scores were calculated by subtracting the PMI value of a term associated with a negative sentiment from the PMI value of a term associated with a positive sentiment [9]. The negated content is distinguished into two groups based on whether the term is immediately following a negated word or whether the term resides within negated context.

LIWC2015 framework is mainly used to identify the impact of standard linguistic constructs such as pronouns, articles, negations, etc. and psychological constructs such as positive and negative emotion, anger, sadness, etc., on the sentiment value imposed on the topic by the given sentence [19].

4.3   Sentence-Level Feature Extraction

Sentence-level feature extraction is conducted mainly using the above-mentioned lexicons. Hashtags in a tokenized Twitter message were looked up in the hashtag emotion lexicon, and the scores were aggregated according to the associated values for each category of emotion. If the given hashtags are not being associated with any of the emotion classes, a value of zero is being returned for the sentence for the specific emotion class.

As an additional attribute, the aggregated emotion values were compared to the maximum value, which is being assigned as the representative nominal class for the given sentence.

In order to compute the features based on the word emotion lexicon, the tokenized Twitter message was matched against the lexicon and the associated values were aggregated according to each individual emotion class in order to represent the sentence.

The MaxDiff Twitter sentiment lexicon is used to identify the aggregated scores for a sentence with the associated values given for unigrams. As the values represent positive sentiment towards a given word calculated using the MaxDiff method of annotation, positive and negative value aggregation has resulted in a representation of a sentiment value for the given tweet.

Also, SentiWordNet is used to obtain an aggregated value for the sentence by matching words between the tokenized tweet and the SentiWordNet synonym sets. In addition to the sentence level SentiWordNet score, the given topic in a message is also being evaluated against the synonym set to identify if it carries a sentiment value.

Tokenized Twitter bigrams are also being used to identify related bigram lexical entries against the "Sentiment140" lexicon. In total, at the message level, the classifier was trained on nine features using the hashtag emotion lexicon, ten features using the word-emotion association lexicon, and one feature each using the MaxDiff Twitter sentiment lexicon and SentiWordNet. Also, the Sentiment140 lexicon for unigrams and bigrams was used in identifying one feature each at the message level.

4.4   Topic-Level Feature Extraction

Topic-level feature extraction is implemented similarly to sentence-level feature extraction using the above-mentioned lexicons. The key motivation behind the identification of the dependent words is the nature of the task, where it is required to identify the sentence polarity towards a given topic. It is noted that the sentence level polarity and the sentence polarity towards a given topic can be different, as the topic might or might not contribute towards the overall polarity of the sentence. Dependency parsing is being used mainly to identify the sentiment contribution made by the dependent tokens towards the topic, as all the tokens within the sentence might not contribute equally towards the sentiment of a sentence. In contrast to the feature extraction based on the associated tokens towards the left and right of the specific topic [9], the dependency token identification can be intuitively considered as an effective methodology due to the following reasons: the neighboring tokens might emphasize less the sentiment value; and, most importantly, the token selection can be limited based on their dependency relation to the topic.

The output obtained from the TweeboParser is analyzed to identify both tokens being dependent on the topic and the relevant dependencies that the topic has with the rest of the tokens within a given sentence. The multi-word topics are considered as units and the dependencies towards and from them are identified. Extracted topic dependencies are evaluated using the given lexicons to identify different attributes, as mentioned above under different lexicon features. The features are identified against both unigrams and bigrams according to the given lexicon.

In total, at the topic level, the classifier was trained on nine features using the hashtag emotion lexicon, ten features using the word-emotion association lexicon, two features using the SentiWordNet and one feature using the MaxDiff Twitter sentiment lexicon. In addition, the Sentiment140 lexicon with unigrams and bigrams was used to identify one feature each at the topic level.

In summary, a total of 47 features were used to train the classifier (23 at message level and 24 at topic level) and considerable improvement was obtained by using both sentence- and topic-level features, as it will be described in the next section.

In addition to the above-mentioned lexicons used at both sentence and topic levels, we have used several other freely available (specifically for academic research purposes) tweet-specific sentiment lexicons (made available by NRC Canada) and a commercial lexicon also available for academic research. The commercial lexicon is Linguistic Inquiry Word Count (LIWC2015 framework) which was designed mainly to identify linguistic and psychological constructs. The use of this lexicon was not limited to the above-mentioned categories; all the other categories were used in order to identify different language constructs.

In addition to the features being identified with the use of above-mentioned lexicons, we calculated the use of punctuation marks, which includes the use of commas, periods, questions marks, etc.

5   Results

The evaluation measure that we report is the one used in the SemEval task: the macro average F1 measure for the positive class and for the negative class (excluding the neutral class). The key reason that could be identified as the motivation behind the use of this macro F1-measure is the unequal distribution of the classes (the neutral class being dominant in the dataset).

Table 2. Comparison of the classification results generated using sentence- and topic-level features together, while removing subsets of features at sentence-level, at topic-level or at both levels.

Features (Lexicons)	Macro F1-measure	
	Sentence level	Topic level
Include all features	0.4845	
Remove all features at one level but keep them for the other level	0.3500	0.4435
Sentiment140 lexicon (bigrams)	0.4680	0.4805
Sentiment140 lexicon (unigrams)	0.4730	0.4945
SentiWordNet	0.4745	0.4825
MaxDiff Twitter sentiment lexicon	0.4845	0.4995
Word-emotion association lexicon	0.4845	0.4845
Hashtag emotion	0.5140	0.4745
Hashtag + Word-emotion + MaxDiff + Sentiment140 lexicon (unigrams) (topic)	0.5230	
Hashtag + Word-emotion + MaxDiff (sentence) + MaxDiff (topic)	**0.5275**	
Hashtag + MaxDiff (sentence) + MaxDiff (topic)	**0.5310**	

As the first step in evaluation, the most efficient and effective machine-learning algorithm to be used as the main classifier was identified as decision trees, compared with the results[2] obtained for different classifiers such as Support Vector Machines (SVM) and Naive Bayes. Decision trees resulted[3] with the highest macro average F1 measure for both positive and the negative classes, given all the feature vectors.

To understand the impact of different features identified through lexicons and the impact sentence- and topic-level features have on the overall classifier performance, we separately ran the decision trees algorithm on sentence- and topic-level features. Through such analysis, as well as by conducting attribute subset evaluation and information gain with respect to the class on separate features at sentence- and topic-level, we could identify that SentiWordNet and Sentiment140 lexicon features have more influence on the classifier performance followed by Word-emotion, MaxDiff and Hashtag emotion lexicons.

Table 2 summarizes the results obtained for different combinations of features, at both sentence- and topic-level. The first line includes all features at

---

[2] Comparing the weighted average F1 measure, the results obtained using a t-test with both sentence- and topic-level features for decision trees (0.64) was noticeably higher than SVM (0.60) and statistically significant than Naive Bayes algorithm (0.44).

[3] Decision trees macro average F1 measure (0.48) was substantially higher than both SVM (0.39) and Naive Bayes (0.35) macro average F1 measure.

both levels. The second line removes all the sentence-level features and keeps only topic-level features in the first column of results and removes all the topic-level features, but keeps the sentence-level features in the second column of results. Then, the next lines remove one or more lexicons at a time from each level, and in the last three lines from both levels.

Via further experiments on the dataset with the use of lexicons such as NRC Hashtag Sentiment, Sentiment140 Affirmative Context, Sentiment140 Negated Context, NRC Hashtag Affirmative Context Sentiment, NRC Hashtag Negated Context Sentiment, LIWC2015 framework, and with the extracted punctuation marks, we have identified that the Support Vector Machine algorithm produces better results compared to decision trees. Even though the results cannot be considered as significant (the macro F1-measure is only slightly higher, at 0.54144), we have identified that with the use of information gain attribute evaluation, that several features introduced through the newly added lexicons (e.g., NRC Hashtag Sentiment Lexicon, NRC Hashtag Affirmative Context Sentiment Lexicon, NRC Hashtag Negated Context Sentiment Lexicon and LIWC categories such as: cognitive processes, affective processes, informal language) have contributed positively towards the obtained accuracy.

6 Discussion

Compared to the best results (macro-F1 scores) obtained by the participants of the SemEval 2015 task 10 subtasks C, our results can be considered equally competitive (though we did not have access to the test data that the participants used, therefore we cannot compare directly). The good results that we obtained were mainly due to the use of the publicly available lexicons and the rich set of lexicons provided by NRC Canada through extensive research on sentiment analysis of short informal text. Both sentence- and topic-level features have contributed to the high accuracy level, while sentence-level features can be identified as the main contributor.

The use of topic level features identified through topic dependencies has provided a substantial improvement in the overall results by increasing the macro-F1 measure from 0.4435 (using only sentence-level features) to 0.5310 (using both sets of features, but with fewer sentence-level features compared to topic-level features).

7 Conclusions and Future Work

The identification of both sentence-level features and topic-level dependencies with the use of lexicons designed especially for short informal texts, such as tweets, have caused our proposed solution to achieve good results. It was also identified that introducing more features based on lexicons at sentence- and topic-level could further increase the accuracy of the classifier. Even though the newly obtained accuracy could not be considered as a significant improvement, we could still see more improvement if we would have more training data.

The dataset that we obtained for this work contained 391 twitter messages out of the 489 initially labeled in the SemEval task, due to unavailability of some of the messages.

In future work, in addition to lexicon-based features, factors that have high impact on sentiment, such as identification of negation, part-of-speech tagging and tag frequencies could also be considered in order to improve the accuracy of the classifier. Further identification of dependency relations by training the dependency parser with additional dependency relation labels, could also improve the accuracy level of the classifier. We also plan to do more extensive testing, on large amounts of tweets that arrive in real time for various target topics.

References

1. Baccianella, S., Esuli, A., Sebastiani, F.: SentiWordNet 3.0: an enhanced lexical resource for sentiment analysis and opinion mining. In: Calzolari, N., Choukri, K., Maegaard, B., Mariani, J., Odijk, J., Piperidis, S., Rosner, M., Tapias, D. (eds.) Proceedings of the Seventh International Conference on Language Resources and Evaluation (LREC 2010), Valletta, Malta, May 2010. European Language Resources Association (ELRA) (2010)
2. Bifet, A., Holmes, G., Pfahringer, B., Gavaldà, R.: Detecting sentiment change in Twitter streaming data. In: 2nd Workshop on Applications of Pattern Analysis, vol. 17, pp. 5–11 (2011)
3. Boag, W., Potash, P., Rumshisky, A.: TwitterHawk: a feature bucket approach to sentiment analysis. In: Proceedings of the 9th International Workshop on Semantic Evaluation (SemEval 2015), pp. 640–646 (2015)
4. Buchholz, S.: CoNLL-X Shared Task: Multi-lingual Dependency Parsing (2006)
5. Fernández, J., Gutiérrez, Y., Gómez, J.M., Patricio Martínez-Barco, G.: Supervised sentiment analysis in Twitter using skipgrams. In: Proceedings of the 8th International Workshop on Semantic Evaluation (SemEval 2014), pp. 294–299 (2014)
6. Filho, P.P.B., Avanco, L., Pardo, T.A.S., Nunes, M.G.V.: NILC_USP: a hybrid system for sentiment analysis in Twitter messages. In: Proceedings of the 8th International Workshop on Semantic Evaluation (SemEval 2014), pp. 428–432 (2014)
7. Jansen, B.J., Zhang, M., Sobel, K., Chowdury, A.: Twitter power: tweets as electronic word of mouth. J. Am. Soc. Inf. Sci. Technol. 60(11), 2169–2188 (2009)
8. Jurafsky, D., Martin, J.H.: Speech and Language Processing, 2nd edn. Prentice-Hall Inc., Upper Saddle River (2009)
9. Kiritchenko, S., Zhu, X., Mohammad, S.M.: Sentiment analysis of short informal texts. J. Artif. Intell. Res. 50, 723–762 (2014)
10. Kong, L., Schneider, N., Swayamdipta, S., Bhatia, A., Dyer, C., Smith, N.A.: A dependency parser for tweets. In: EMNLP 2014 (2014)
11. Kouloumpis, E., Wilson, T., Moore, J.: Twitter sentiment analysis: the good the bad and the OMG! In: Proceedings of the Fifth International AAAI Conference on Weblogs and Social Media (ICWSM 2011), pp. 538–541 (2011)
12. Manning, C.D., Jurafsky, D.: Sentiment Analysis (2015)
13. Mohammad, S.M., Kiritchenko, S.: Using hashtags to capture fine emotion categories from tweets. Comput. Intell. 31(2), 301–326 (2015)

14. Mohammad, S.M., Kiritchenko, S., Zhu, X.: NRC-Canada: building the state-of-the-art in sentiment analysis of tweets. CoRR, abs/1308.6 (SemEval), pp. 321–327 (2013)
15. Mohammad, S.M., Turney, P.D.: Crowdsourcing a word-emotion association lexicon. Comput. Intell. **59**(000), 1–24 (2011)
16. Owoputi, O., O'Connor, B., Dyer, C., Gimpel, K., Schneider, N., Smith, N.A.: Improved part-of-speech tagging for online conversational text with word clusters. In: Proceedings of NAACL-HLT 2013, pp. 380–390, June 2013
17. Pak, A., Paroubek, P.: Twitter based system: using Twitter for disambiguating sentiment ambiguous adjectives. In: Proceedings of the 5th International Workshop on Semantic Evaluation, pp. 436–439, July 2010
18. Pang, B., Lee, L.: Opinion mining and sentiment analysis. Found. Trends Inf. Retrieval **2**(1–2), 1–135 (2008)
19. Pennebaker, J.W., Boyd, R., Jordan, K., Blackburn, K.: The development and psychometric properties of LIWC2015 (2015)
20. Plotnikova, N., Kohl, M., Volkert, K., Lerner, A., Dykes, N., Ermer, H., Evert, S.: KLUEless: polarity classification and association. In: Proceedings of the 9th International Workshop on Semantic Evaluation (SemEval 2015), vol. 1, pp. 619–625 (2015)
21. Quinlan, J.R.: Induction of decision trees. Mach. Learn. **1**(1), 81–106 (1986)
22. Rosenthal, S., Nakov, P., Kiritchenko, S., Mohammad, S., Ritter, A., Stoyanov, A: SemEval-2015 Task 10: sentiment analysis in Twitter. In: Proceedings of the 9th International Workshop on Semantic Evaluation, SemEval, pp. 451–463 (2015)
23. Scherer, K.R.: Emotion as a multicomponent process: a model and some cross-cultural data. Rev. Pers. Soc. Psychol. **5**, 37–63 (1984)
24. SemEval-2015. SemEval-2015 (2015)
25. Townsend, R., Tsakalidis, A., Wang, B., Liakata, M., Cristea, A., Procter, R.: WarwickDCS: from phrase-based to target-specific sentiment recognition. In: Proceedings of the 9th International Workshop on Semantic Evaluation (SemEval 2015), pp. 657–663 (2015)
26. Villena-román, J., García-Morera, J., González-cristóbal, J.C., García-morera, J.: DAEDALUS at SemEval-2014 Task 9: comparing approaches for sentiment analysis in Twitter. In: Proceedings of the 8th International Workshop on Semantic Evaluation (SemEval 2014), pp. 218–222 (2014)
27. Bin Wasi, S., Neyaz, R., Bouamor, H., Mohit, B.: CMUQ@Qatar: using rich lexical features for sentiment analysis on Twitter. In: Proceedings of the 8th International Workshop on Semantic Evaluation (SemEval), pp. 186–191 (2014)
28. Zhang, Z., Guoshun, W., Man Lan, E.: Multi-level sentiment analysis on Twitter using traditional linguistic features and word embedding features. In: Proceedings of the 9th International Workshop on Semantic Evaluation (SemEval 2015), pp. 561–567 (2015)
29. Zhao, J., Lan, M., Zhu, T.T.: ECNU: expression- and message-level sentiment orientation classification in Twitter using multiple effective features. In: Proceedings of the 8th International Workshop on Semantic Evaluation (SemEval 2014), pp. 259–264 (2014)

A Security Price Data Cleaning Technique: Reynold's Decomposition Approach

Rachel V. Mok[1], Wai Yin Mok[2(✉)], and Kit Yee Cheung[2]

[1] Department of Mechanical Engineering,
Massachusetts Institute of Technology, Cambridge, MA 02139, USA
rmok@mit.edu
[2] College of Business Administration, University of Alabama in Huntsville,
Huntsville, AL 35899, USA
mokw@uah.edu, kityeemok@gmail.com

Abstract. We propose a security price data cleaning technique based on Reynold's decomposition that uses T_I, the time period of integration, to determine the de-noise level of the price data. As price is a function of time, T_0, the optimal time period of integration, may reveal an underlying price trend, possibly indicating the intrinsic value of the security. The DJIA (Dow Jones Industrial Average) Index and the thirty companies comprising the index are our fundamental interest under the initial investigation period from 1990 to 2016. Also, intra-day security price data from February 8th to August 19th, 2016 are obtained to further study T_0 on a minute-by-minute basis. Preliminary results include the following: (1) It was discovered that α, a key percentage measure, drops exponentially for low T_I and then drops linearly at a fairly shallow slope for high T_I. (2) In the linear region, the α hardly varies as T_I increases. Thus, we propose that the optimal time period of integration, T_0, is when α transitions from an exponential behavior to a linear behavior. We calculated that the average of the T_0's for the thirty DJIA component companies is 64 business days and that for the DJIA itself is 63 business days. For intra-day study of T_0, α seems to drop proportionally with the length of T_I, exhibiting an almost linear relationship. The change in slope for the intra-day study is not as noticeable as the total time period study. The average of the intra-day T_0's for the thirty DJIA component companies is 52 min and for the DJIA Index is 69 min.

1 Introduction

Understanding and analyzing financial data in order to forecast and make cost-effective decisions is challenging because of the complex and volatile nature of security prices. The most recent financial market meltdown in 2008–2009 casted doubts on financial data analysis and forecasting. Inability to recognize or acknowledge financial distress signaled by pertinent financial data was a significant factor leading to these catastrophic economic results [9]. Thus, veracity of financial data takes priority in any data driven decision making. Like any big

© Springer International Publishing AG 2017
J.A. Lossio-Ventura and H. Alatrista-Salas (Eds.): SIMBig 2015/2016, CCIS 656, pp. 108–119, 2017.
DOI: 10.1007/978-3-319-55209-5_9

data infrastructure, veracity includes validation, noise level, deception, detection, relevance and ranking of data collected [6]. Depending on how collected financial data are captured and processed in an analysis, generated assessments can vary greatly from real financial market performance. One has to look no farther than the recent settlement of $77 million between the SEC and Standard & Poor credit rating agency to see an example of how data analysis can be misleading (http://www.sec.gov/news/pressrelease/2015-10.html).

Several financial computation models that deal with cleaning financial data employ similar methodologies, such as candlestick strategies [5], multiple-stage algorithm for detecting outliers in ultra high-frequency financial market data [20], financial data filtering (http://www.olsendata.com) and data-cleaning algorithm [2,3]. Most data cleaning methodologies involve the detection, distribution and/or the removal of outliers [16,18]. However removing outliers in the dataset may have a statistical distortion effect on the dataset itself [4].

To this end, we propose a data cleaning technique based on Reynold's decomposition in order to decompose the price data into a mean part and a fluctuating part. Fluctuations in stock prices are perpetual and irrational in time because the weak form of market efficiency and different types of market participants create a complex dynamic of behavioral finance [19]. Nevertheless, our approach could minimize part of the effect of irrational price fluctuations by incorporating and averaging fluctuation points (i.e., outliers) within a moving time period of integration, T_I. In essence, the length of T_I in the analysis determines the level of veracity, with the larger the T_I, the lesser the influence of the fluctuation points will be. We believe our data cleaning technique is particularly applicable to security prices due to the intense nature of security price changes in relatively short periods, and it allows the user to gauge different moving time periods of integration to produce a unique set of statistical data for targeted analysis.

2 Reynold's Decomposition

In the study of turbulence in fluid dynamics, each component of the velocity is characterized by fluctuations over time. One method to study the dynamics in this regime is to perform a Reynold's decomposition such that the mean part of the velocity is separated from the fluctuations. We propose that this technique could also be used to study financial data. In other words, we propose that the price as a function of time, $p(t)$, can be decomposed into the following:

$$p(t) = \bar{p}(t) + p'(t) \tag{1}$$

where $\bar{p}(t)$ is the mean portion and $p'(t)$ is the fluctuating portion of the price. We define $\bar{p}(t)$ to be a moving time-average that can be found by performing the following integral

$$\bar{p}(t) = \frac{1}{T_I} \int_{t-T_I/2}^{t+T_I/2} p(t')dt' \tag{2}$$

where T_I is the time period of integration. T_I must be a time period that is greater than the time period of the fluctuations, τ, and less than the time period of interest, T. T is dependent on each particular analysis; for example, T could be weeks, months, or years. Thus, $\tau < T_I < T$. Furthermore, the time-averaged value of the fluctuating portion over the entire time period of interest is zero [13,15]. As the time period of integration increases, $\bar{p}(t)$ is tends to be farther away from the actual $p(t)$ and the magnitude of $p'(t)$ increases. Thus, the goal of this research is to find the optimal time period of integration, T_0, that excludes the miscellaneous fluctuations and captures the essential trend of the price data.

3   Methods

In this study, we focus on the thirty companies comprising the Dow Jones Industrial Average (DJIA) as of September 2, 2016, and the DJIA Index because, being the second oldest financial index, the DJIA is the benchmark that tracks financial market performance as a whole. Thus, it represents a broad market, and its validity is intensely scrutinized and followed by at least 10 Wall Street analysts [10,14,17]. The ticker symbols for the thirty companies that were studied in this analysis are as follows: AAPL, AXP, BA, CAT, CSCO, CVX, DD, DIS, GE, GS, HD, IBM, INTC, JNJ, JPM, KO, MCD, MMM, MRK, MSFT, NKE, PFE, PG, TRV, UNH, UTX, V, VZ, WMT, and XOM. Because different companies can comprise the DJIA Index at any point in time, we only focus on the index as a whole when performing the analysis for the DJIA Index itself.

Daily last price stock price data for the thirty Dow Jones companies listed above are obtained from January 2, 1990, until September 2, 2016, from the Bloomberg Professional Service. If the company's date of inception was after January 2, 1990, then the date of inception was used as the start date. For the DJIA Index, the daily last stock price data from January 2, 1990, to September 2, 2016, are also obtained from the Bloomberg Professional service. Only days in which the stock price is provided, i.e., business days, are considered in this study. Thus, the time from Friday to Monday is taken as only one (business) day.

We estimate the time period of fluctuations to be a day, $\tau \sim 1$ business day, and the time period of interest to be the total number of business days for each time period of analysis. In other words, $T = 691$ days, $T = 376$ days, $T = 502$ days, $T = 1365$ days, and $T = 6702$ days for the peak, contraction, expansion, recovery, and total time periods, respectively. Further, we chose the following time periods of integration, T_I, for this study: 5 days, 9 days, 19 days, 41 days, 61 days, 121 days, 181 days, 261 days, which roughly represent the following time periods: one week, two weeks, one month, two months, three months, half of a year, three-quarters of a year, and one year, respectively.

$\bar{p}(t)$ is calculated by only considering the time period from $T_I/2 - 0.5$ after the start date of each time period of analysis to $T_I/2 + 0.5$ before the end date of each time period of analysis, such that for each day $\bar{p}(t)$ is calculated, the full time period of integration is used. To exemplify, consider the case where $T = 10$ days and $T_I = 5$ days. Then the first 2 days (day 1 to day 2) are not included in the analysis, and neither are the last 2 days (day 9 to day 10). For each day in the analysis time period, the integration stated in Eq. (2) is performed numerically to find $\bar{p}(t)$ for that day. $p'(t)$ is found by subtracting $\bar{p}(t)$ from $p(t)$, the actual price, for that day.

For each specific T_I, the statistics of $p'(t)$ are analyzed. Specifically, a histogram of $p'(t)$ is created for each T_I. As an example, Fig. 1 shows the histograms for GE (General Electric Corporation) for the overall time period of analysis that spanned from January 2, 1990, to September 2, 2016. The histograms shown in Fig. 1 exhibit general characteristics seen in many of the other companies. Specifically, the largest number of observations of $p'(t)$ for each T_I is usually at zero, which suggests that most of the fluctuations for the stocks are nearly zero. Therefore, the actual stock price is near or nearly equal to the local time-average for most of the time period analyzed. Furthermore, many of the companies' histograms are symmetric about zero, which suggest that the local time-average calculated is truly the average of the actual price. Also, as T_I increases, the maximum height of the histogram decreases and the histogram tails become heavier. Thus, as T_I increases, there are more observations away from the center of the distribution. This is observed because as the time period of integration increases, more points are considered in the average. Therefore, there is a greater likelihood that $\bar{p}(t)$ is different from the actual price.

To measure the fidelity of $\bar{p}(t)$ to $p(t)$, the number of data points of $p'(t)$ that are within 1% of the entire fluctuation range at the largest T_I considered from zero are counted and divided by the total number of data points in the analysis period for each T_I. We will call this percentage measure α, and this measure should be as close as possible to 100% to reflect that $\bar{p}(t)$ is a good approximation of $p(t)$. As stated previously, if $p'(t)$ is near zero, that means that $\bar{p}(t)$ is close to $p(t)$ because $p(t) = \bar{p}(t) + p'(t)$. As T_I increases, α decreases because the mean is farther away from the actual price when the integration period is larger. Using the MATLAB® curve fitting tool, it is found that for all of the thirty stocks the relationship between α and T_I is best represented by the following equation

$$\alpha = a \, \exp\left(b \, T_I\right) + c \, \exp\left(d \, T_I\right) \tag{3}$$

where a, b, c, and d are curve fitting parameters. Shown in Fig. 2 are the α at each analyzed T_I considered for GE during the overall time period and the fitted *alpha* curve.

For most of the stocks, it was discovered that α drops exponentially for low T_I and then drops linearly at a fairly shallow slope for high T_I as exhibited by Fig. 2. In the linear region, the α hardly varies as T_I increases. Thus, after a certain time period of integration, the fidelity of $\bar{p}(t)$ to $p(t)$ is nearly equal for all T_I considered. Thus, we propose that the optimal time period of integration, T_0, is when α transitions from an exponential behavior to a linear behavior. To find the point of transition, we can take the first and second derivative of Eq. (3) which are, respectively,

$$\frac{d\alpha}{dT_I} = a \, b \, \exp\left(b \, T_I\right) + c \, d \, \exp\left(d \, T_I\right) \tag{4}$$

and

$$\frac{d^2\alpha}{dT_I^2} = a \, (b)^2 \, \exp\left(b \, T_I\right) + c \, (d)^2 \, \exp\left(d \, T_I\right) \tag{5}$$

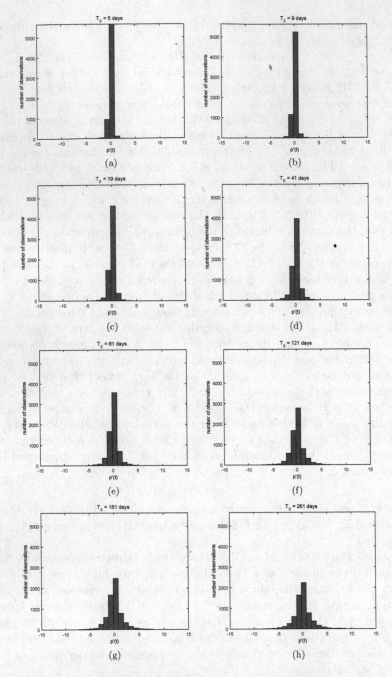

Fig. 1. Histograms of $p'(t)$ for each T_I analyzed for GE (General Electric Corporation) from January 2, 1990, to September 2, 2016, (i.e. the overall time period of analysis).

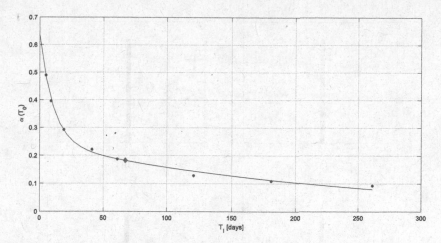

Fig. 2. Plot of α as T_I varies for GE during the overall time period. The blue dots are the calculated α for each of the T_I considered in this study. The black line is the fitted curve found using MATLAB® curve fitting tool where $a = 0.3952, b = -0.09287, c = 0.2411$, and $d = 0.004252$. The red diamond is T_0, the calculated optimal time period of integration. Note that at $T_0 = 67$ days, the fitted curve changes from an exponential behavior to a linear behavior. (Color figure online)

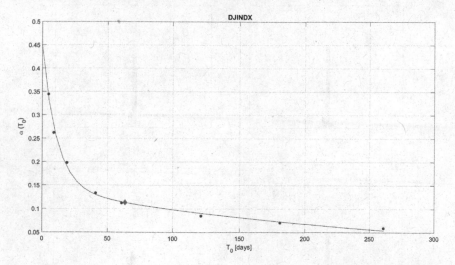

Fig. 3. T_0 on daily data for DJIA Composite Index from January 2, 1990, to September 2, 2016.

and find the first point at which Eq. (5) becomes zero or nearly zero. As an example, for GE during the overall time period, T_0 was calculated to be 67 days. As shown in Fig. 2, where T_0 is represented as a red dimond, for $T_I < T_0$, the fitted α curve exhibits an exponential behavior, but for $T_I \geq T_0$, the fitted α curve exhibits a linear behavior (Fig. 3).

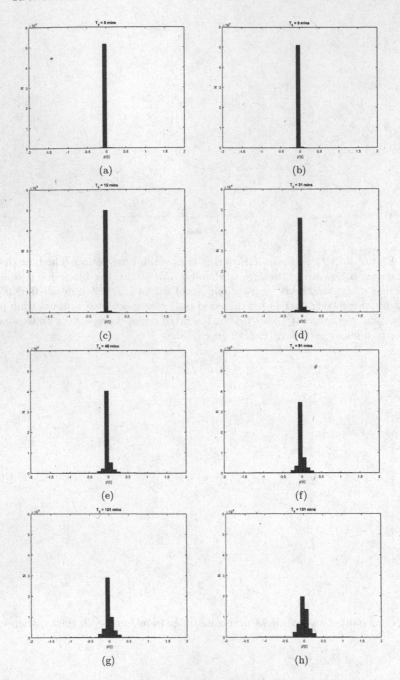

Fig. 4. Histograms of $p'(t)$ on minute-by-minute data for each T_I analyzed for GE (General Electric Corporation) from February 8th to August 19th, 2016.

To extend our daily data study on T_0, minute-by-minute stock price data of the thirty DJIA component stocks and the DJIA composite index were considered and collected from February 8th to August 19th, 2016. However, due to the large volume of the data and the restriction of the Bloomberg Professional Service, we were only able to collect the last six months of the intraday data from February 8th to August 19th, 2016 (Fig. 4).

4   Results and Conclusions

As a whole, regarding the time period of integration for each of the DJIA thirty component companies and the DJIA Composite Index, we found that the graphs of α versus T_I of the thirty component companies and the DJIA Index exhibit at least two properties: (1) α drops exponentially as T_I increases when $T_I \lessapprox$ order of magnitude of 64 business days, and (2) α drops linearly as T_I increases when $T_I \gtrapprox$ order of magnitude of 64 business days. Thus, the optimal T_0 for the thirty component companies is approximately 64 business days. For the DJIA Index itself, the optimal T_0 is 63 business days, reflecting the average T_0 of the DJIA thirty component companies. Thus, the two measures are almost identical. One of the possible explanations is that the DJIA Index might show the counter measure effect of fluctuation points among the thirty companies since the DJIA is a composite of the thirty companies that collectively provide a balance view of the market. As a result, T_0 might be even smaller for the second derivative to approach zero. We also found that σ and T_0 are related by a power law.

As shown in Table 1, the optimal time period of integration for each of the DJIA thirty component companies ranges from 44 to 77 business days and averages 64 business days, which is equivalent to 90 calendar days for the period from 1990 to 2016; and for more than half of the thirty component companies, T_0 falls in the range from 58 to 67 business days. For DJIA Composite Index itself, T_0 is 63 business days, which is equivalent to 88 calendar days. The optimal time periods of integration for the DJIA thirty component companies and the DJIA Composite Index itself coincide with the SEC required filing periods of 10 K, which is a quarterly report filed every 90 calendar days by public companies including unaudited financial statement, quarterly earnings and forward looking statements. Studies have suggested that 10 K quarterly reports may closely relate to changes in stock prices [1, 7, 8, 11, 12]

At a closer look, IBM has the longest T_0 of 77 business days, followed by Chervon's T_0 of 73 business days, and followed by the T_0 of 72 business days for Nike, Pfizer, United Health Group Inc, and United Technologies Corporation. On the contrary, Apple has the shortest T_0 of 44 business days, followed by both Visa's T_0 and Verizon's T_0 of 52 business days. In fact, the longer the T_0, the further away is the fluctuating part from the moving average part of the stock price, meaning that the longer T_0 contains relatively more data quantities than the average T_0 to approximate the actual price. The opposite is true with the shorter T_0. However, the length of T_0 does not indicate the fluctuation magnitude compared to the actual price.

Table 1. T_0's for each DJIA component company for both daily data and minute-by-minute data.

Company	Total (business days)	Intraday (minutes)
AAPL	44	53
AXP	71	52
BA	64	53
CAT	69	53
CSCO	66	50
CVX	73	53
DD	56	52
DIS	64	53
GE	67	52
GS	52	52
HD	62	53
IBM	77	53
INTC	68	50
JNJ	66	53
JPM	65	53
KO	61	51
MCD	56	53
MMM	58	51
MRK	70	53
MSFT	66	53
NKE	72	53
PFE	72	53
PG	60	53
TRV	61	50
UNH	72	53
UTX	72	53
V	52	53
VZ	52	52
WMT	68	52
XOM	64	53
Average	**64**	**52**
DJIA Index	63	69

Since the definition of T_0 is that when α transitions from an exponential behavior to a linear behavior, then the longer the T_0, the less L-shape of the curve will exhibit. It may suggest that longer T_0's have less frequent price changes, which may either be increasing or decreasing. Although it seems that longer T_0's

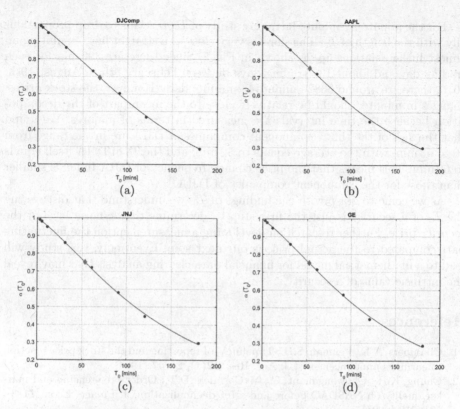

Fig. 5. Four different plots of α as T_I vary for intra-day study of DJComp, APPL, JNJ, and GE from February 8th to August 19th, 2016. In each individual graph, the blue dots are the calculated α for each of the T_I considered in this study, the black line is the fitted curve found using MATLAB® curve fitting tool, and the red diamond is T_0, the calculated optimal time period of integration. Note that T_0's are 69, 53, 53, 52 respectively for DJComp, APPL, JNJ, and GE. (Color figure online)

have less frequent price changes, the relationship between the length of T_0 and the fluctuation magnitude is inconclusive. It is then possible that T_0's of different length may yield a similar fluctuation magnitude.

Recall our understanding that the length of T_0 indicates the frequency of price changes, but not the fluctuation magnitude. The apparent lack of correspondence between Beta, a measure of financial market volatility or systematic risk of a security, and a particular stock's T_0 provides another anecdotal evidence that the relationship between the stock's fluctuation magnitude and the length of its T_0 is inconclusive. For example, Verizon has the lowest Beta of 0.26, meaning that it is relatively stable compared to the market as a whole. However, its T_0 is relatively short, meaning that it has a relatively higher frequency of price changes. Similar observations can be found in Nike (Beta = 0.5), United Technologies Corporation (Beta = 0.57), IBM (Beta = 0.77), Pfizer (Beta = 0.87), Visa (Beta = 0.98), United Health Group (Beta = 1.06), Apple (Beta = 1.16) and Chevron (Beta = 1.17).

For the minute-by-minute intra-day study of T_0, α seems to drop proportionally with the length of T_I. The slope is very close to a straight line, exhibiting an almost linear relationship, as shown in Fig. 5. Since there are 390 min for each trading day and about 138 trading days between February 8th to August 19th, 2016, there are around 53820 minute-by-minute data. However, the stock prices' changes in minutes should be relatively close to the mean part of the decomposition because the time intervals are measured in terms of minutes. We found that the T_0's of the thirty component companies of DJIA are in the range from 50 to 53 min, with the average equals to 52 min; and the T_0 of DJIA itself equals to 69 min. This means that the price change frequency of DJIA Index is smaller than those for the component companies of DJIA.

As we come to grasp with the findings of T_0, we understand that discovering the T_0 of a security is just the first step to determine the de-noise level of the security price. Further research on developing a measurement on the fluctuating part compared to the actual price is our next step. Eventually, this study will lead to a mathematical model for financial data clearing analysis that may reveal the intrinsic value of a security.

References

1. Chambers, A.E., Penman, S.H.: Timeliness of reporting and the stock price reaction to earnings announcements. J. Acc. Res. **22**(1), 21–47 (1984)
2. Chung, K.H., Chuwonganant, C., McCormick, D.T.: Order preferencing and market quality on NASDAQ before and after decimalization. J. Financ. Econ. **71**(3), 581–612 (2004)
3. Chung, K.H., Van Ness, B.F., Van Ness, R.A.: Trading costs and quote clustering on the NYSE and NASDAQ after decimalization. J. Finan. Res. **27**(3), 309–328 (2004)
4. Dasu, T., Loh, J.M.: Statistical distortion: consequences of data cleaning. Proc. VLDB Endowment **5**(11), 1674–1683 (2012). http://vldb.org/pvldb/pvldb/vol5/p1674_tamraparnidasu_vldb2012.pdf
5. Detollenaere, B., Mazza, P.: Do Japanese candlesticks help solve the trader's dilemma? J. Bank. Finance **48**, 386–395 (2014)
6. Goes, P.B.: Big data and IS research. MIS Q. **38**(3), iii–viii (2014)
7. Jordan, R.J.: An empirical investigation of the adjustment of stock prices to new quarterly earnings information. J. Financ. Quant. Anal. **8**(4), 609–620 (1973)
8. Joy, O.M., Litzenberger, R.H., McEnally, R.W.: The adjustment of stock prices to announcements of unanticipated changes in quarterly earnings. J. Acc. Res. **15**(2), 207–225 (1977)
9. Kaur, I.: Early warning system of currency crisis: insights from global financial crisis 2008. IUP J. Appl. Econ. **14**(1), 69–83 (2015)
10. Lee, C.M.C., Swaminathan, B.: Valuing the dow: a bottom-up approach. Financ. Anal. J. **55**(5), 4–23 (1999)
11. Lee, Y.J.: The effect of quarterly report readability on information efficiency of stock prices. Contemp. Acc. Res. **29**(4), 1137–1170 (2012)
12. Mashruwala, C., Mashruwala, S.: Is there a "torpedo effect" in earnings announcement returns? The role of short-sales constraints and investor disagreement. J. Acc. Auditing Finance **29**(4), 519–546 (2014)

13. Mills, A.F.: Basic Heat and Mass Transfer, 2nd edn. Prentice Hall, Upper Saddle River (1999)
14. Moroney, R.: What we're thinking add it up: dow has further upside. Dow Theory Forecasts **68**(9), 2–3 (2012)
15. Müller, P.: The Equations of Oceanic Motions. Cambridge University Press, Cambridge (2006)
16. Shamsipour, M., Farzadfar, F., Gohari, K., Parsaeian, M., Amini, H., Rabiei, K., Hassanvand, M.S., Navidi, I., Fotouhi, A., Naddafi, K., Sarrafzadegan, N., Mansouri, A., Mesdaghinia, A., Larijani, B., Yunesian, M.: A framework for exploration and cleaning of environmental data - Tehran air quality data experience. Arch. Iran. Med. **17**(12), 821–829 (2014)
17. Stillman, R.J.: Dow Jones Industrial Average: History and Role in an Investment Strategy. Irwin Professional Pub (1986)
18. Sun, W., Whelan, B., McBratney, A., Minasny, B.: An integrated framework for software to provide yield data cleaning and estimation of an opportunity index for site-specific crop management. Precision Agric. **14**(4), 376–391 (2013)
19. Verheyden, T., De Moor, L., Van den Bossche, F.: Towards a new framework on efficient markets. Res. Int. Bus. Finance **34**, 294–308 (2015)
20. Verousis, T., ap Gwilym, O.: An improved algorithm for cleaning ultra high-frequency data. J. Deriv. Hedge Funds **15**(4), 323–340 (2010)

Big Data Architecture for Predicting Churn Risk in Mobile Phone Companies

Alonso Raul Melgarejo Galvan$^{(\boxtimes)}$ and Katerine Rocio Clavo Navarro

Faculty of Systems Engineering, National University of San Marcos, Lima, Peru
alonsoraulmgs@gmail.com, perclavo@gmail.com

Abstract. Nowadays in Peru, mobile phone companies have been affected by the problem of mobile number portability because since July 2014 customers can change their mobile operator in just 24 h. Companies look for solutions through the analysis of historical data of their customers in order to generate predictive models and to identify which customers would leave the company. However, the current way how this prediction is performed is too slow. In this paper, we show a Big Data architecture which solves the problems of the "classic architecture" using data from social networks in order to predict which customers may go to the competition company, according to their opinions. Data processing is performed by Hadoop, which implements MapReduce and can process large amounts of data in parallel way. After doing the tests and seeing the results, we got a high percentage of accuracy (90.03% of success).

Keywords: Telecommunications · Number portability · Customer churn · Big data · Sentiment analysis · Social network · Naive Bayes · Hadoop · Hive · Mahout

1 Introduction

In the telecommunications area, customer churn is a problem that is becoming increasingly necessary to study due to the high competitiveness that is developing worldwide. The study of Churn or customer leakage is an area in which more resources are invested each year: specialists, external consultants, specialized software, etc. with the intention of discovering in advance, which customers are most at risk of changing from company to the competition [1].

In Peru, customer churn in mobile telephony industry is a problem that currently aspects big telecommunications companies in the country due to the strong competition that has been generated in the market for mobile voice and data services, and which also has generated great commercial offers and price war. Since the entry of mobile operators Bitel and Entel, the competition has increased, being the operator Entel who has obtained more customers from Claro and Movistar Peru [2].

As shown in Fig. 1, according to the latest numbers of [2], it shows that in March, mobile number portability grew 46%, achieving a record of 65,142

© Springer International Publishing AG 2017
J.A. Lossio-Ventura and H. Alatrista-Salas (Eds.): SIMBig 2015/2016, CCIS 656, pp. 120–132, 2017.
DOI: 10.1007/978-3-319-55209-5_10

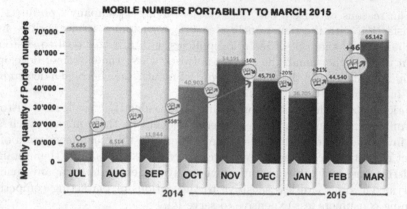

Fig. 1. Mobile number portability to March 2015.

portabilities, the highest since July of last year, date in which the mechanism of number portability was modified and allows to change mobile operator in just 24 h keeping the same phone number.

Therefore, in this paper, we design an Big Data architecture which uses opinion data wrote on Facebook fan page of Claro Peru mobile operator in order to predict which customers may go to the competition company, according to their opinions. Claro is part of the Mexican multinational America Movil and has been operating in Peru since 2005 providing mobile, fixed and internet services. The results obtained in this article were used by the company so that they can generate customer retention strategies.

2 The Issue of Mobile Number Portability: The Customer Churn

Within the telecommunications sector, the word churn or customer churn is used to describe the cessation of subscription services of a customer, and the word churner or absconded refers to a customer who has left the company [1].

A customer can give up the company and to start the termination of his service contract (voluntary churn), or the company may expel him for fraud, non-payment or under-utilization of signed services (involuntary churn). Customer churn can be very costly for a company, because the customer leaks to competition and therefore, not only the unperceived incomes are lost, but also the prestige of the company showed in market participation of competition [3].

2.1 Causes of Customer Churn

The factors which contribute to the leak in telecommunication services are diverse.

Main reasons because a customer ceases buying a company's products are: dissatisfaction with the signal quality, coverage, customer service, prices, irregular charges and the lack of retention policies with a better deal to customers [1], but on the other hand, the contact network where the customer develops is very important because the effect of "word of mouth" has become a determining factor in customers' buying decisions [4].

Anticipating to this problem and being able to identify what leads a customer to terminate his contract, deliver several benefits to the company such as a lower investment in retaining a customer (thanks to the recommendation of the contact network). Acquiring a new customer costs six times more than retaining an existing one, and those customers who stay longer in the company generate higher incomes and are less sensitive to the actions of marketing competition becoming consumers less expensive to serve [4].

2.2   The Current Solution

Nowadays, mobile phone companies develop predictive models in order to identify customers who may escape to competition. The construction of these models could vary, but in general, the steps followed are shown in Fig. 2.

At first, you must identify the data sources with which the prediction model will be built, which correspond to internal data of the company concerning customer profile and call traffic [3].

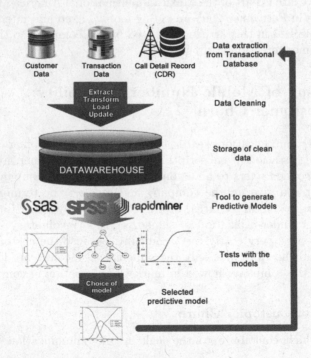

Fig. 2. Steps in the development of predictive models.

The customer profile describes the demographic group of customers and the features they have in common regarding to the segment, the contracted plan, payment type, credit risk, the demographic area and the penalties for not paying the bills. Call traffic describes the traffic generated and received by a customer, who is characterized by calls made and received from an fixed or mobile telephone, local or international calls, from which operator the call was performed, the number of SMS sent and received and the internet traffic generated.

After identifying the sources of data with which we will work, must be performed extraction and cleaning of them. Data cleaning consists in finding errors in data, duplication, inconsistencies or incomplete and inconsistent values. This information is corrected by replacing values generated, or deleted in the worst case. All these data are stored in a data warehouse.

Once the data has been cleaned and are ready for use, different data mining algorithms existing as Naive Bayes, Decision trees or Bayesian network are used to build predictive models. When models are generated, we perform tests on them to find the one with the best ratio of prediction, which will be the model used to predict the number portability.

To measure the effectiveness of models, we measure their accuracy. As [5], mentions, the accuracy of a classifier C is the probability of correctly classifying a randomly selected instance, i.e., acc = $Pr(C(v) = y)$ for a randomly selected instance $(v, y) \epsilon$ X, where the probability distribution over the instance space is the same as the distribution over the instance space is the same as the distribution that was used to select instances for the inducer's training set. Given a finite dataset, we would like to estimate the future performance of a classifier induced by the given inducer and dataset. A single accuracy estimate is usually meaningless without a confidence interval; thus we will consider how to approximate such an interval when possible. In order to identify weaknesses, we also attempt to identify cases where the estimates fail.

On the other hand, as [6] mentions, traditional sequential online algorithms are limited by the memory and bandwidth of a single machine. Distributed stream processing engines (DSPEs) are a family of MapReduce-inspired technologies that address this problem of traditional data mining models.

Also, the customer behavior may change over time, so it is necessary to update the model repeating all the steps.

Now that we understand how prediction models are currently generated, let's see what problems they have.

2.3 Problems of the Existing Solution

Given the current environment in which mobile phone companies develop, the solution currently used to predict customer churn has four problems:

P1. Fast number portability, slow prediction: The new law allows customers can change their mobile operator in just 24 h, so it's necessary to predict the portability quickly in order to avoid it. The classic solution requires maintaining a data warehouse with clean and filtered data, which consumes too much time [2].

P2. Data Confidentiality: Due to telecommunications business confidentiality and customer privacy, it is difficult to find public datasets on churn prediction and thus there is a challenge of standardizing the feature sets to use. The classic solution of prediction works mainly with internal data of the company, because the internal data are structured and available [3].

P3. Unstandardized and inconsistent data: Standardize the characteristics of the different sources of data used for an analysis is a challenge, because it consumes too much time and effort. It's necessary to implement and maintain a data warehouse for data analysis, because this involves removing irrelevant information, duplicated or null values [3].

P4. Changing opinion: The change of customer opinion about a service is not analyzed and this is a determining factor in their decisions [4]. In order to analyze the change of the customer opinions, additional sources to customer profiles and call traffic data should be included.

These four problems cause a slow and inaccurate identification of churn customers, because this requires a cleaning of data or an analysis and sorting of their sources, which shows that the implementation of a traditional architecture to predict number portability is lacking and must be resolved with a different approach.

3   A Different Approach: Big Data

The need to process large volumes of data almost in real time is crucial for businesses today, for example, processing massive volumes of data where valuable information is hidden about the customer buying behavior, and the need to generate new products by analyzing these behaviors. This latter is particularly true in the telecommunications business market, where the number of customers usually reaches several million [1].

This is the scenario where Big Data participates, which analyzes large amounts of data and maximize business value by supporting decision making in business almost in real time [7]. On the other hand, the invention of new technologies has led to the use of large amounts of data which are increasing, also it has created the need to store, process and extract this data. Big Data is responsable of this [8].

3.1   Hadoop Platform

Hadoop is an open-source framework for storing and processing large amounts of data on clusters that work on a commodity hardware [9]. A typical Big Data platform based on Hadoop includes the distributed system of HDFS files, MapReduce framework of parallel computing, a high-level system for managing data such as Pig or Hive, and Mahout as the module for analyzing data [10]. All this is shown in Fig. 3.

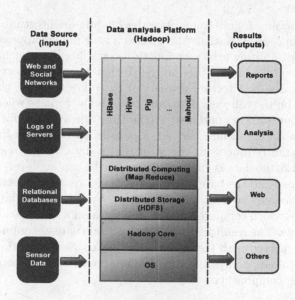

Fig. 3. Hadoop Platform.

HDFS is basically a master/slave structure. Master is known as "Node Name" and slaves as "Data Nodes". The main work of "Node Name" is to store metadata of data, this includes the location of files that contain them and different attributes of documents. The "Data nodes" are responsible for storing data in blocks on different nodes of cluster [8].

MapReduce is a module that Hadoop incorporates for data parallel processing. In order that programmers can write programs on Hadoop, they must specify parameters in the Map Reduce functions in order that Hadoop can run them. Hadoop divides input in many small fragments which are distributed and processed in parallel form on nodes of cluster using the Map function, later through the Reduce function the results are combinated [10].

Hive is a Hadoop module which supports file management on HDFS through a syntax similar to SQL: "Hive Query Language". Hive queries use MapReduce module of Hadoop to run in parallel form and to hide its complexity to programmer. Thanks to this, it's possible to use simple statements on files located in HDFS as "CREATE TABLE", "INSERT", "UPDATE", "SELECT", "DELETE", "JOIN", "GROUP BY" and other valid sentences in the classic SQL [11].

Mahout has been designed for data mining purposes within Hadoop, it implements common algorithms of clustering and regression using MapReduce [12]. Mahout provides tools to automatically find patterns in large volumes of data and it makes an easier and faster analysis within Big Data [9].

As [13] mentions, a lot of technologies can be used to create a realtime processing system. It is complicated to choose the right tools, incorporate, and orchestrate them. Marz proposes a generic solution to this problem, called

Lambda Architecture. Figure 1 describes the lambda architecture. The Lambda Architecture proposed by Nathan Marz takes a very unique approach and solves the problem of computing arbitrary functions on a big data set in real-time by decomposing the problem into three layers: the batch layer, the serving layer, and the speed layer. The architecture consists of three layers. First, the batch layer computes views on the collected data, and repeats the process when it is done to infinity. Its output is always outdated by the time it is available for new data has been received in the meantime. Second, a parallel speed-processing layer closes this gap by constantly processing the most recent data in near realtime fashion. Any query against the data is answered through the serving layer, i.e., by querying both the speed and the batch layers' serving stores in the serving layer, and the results are merged to answer user queries. Speed layer is implemented by Storm (Trident), which computes ad-hoc functions on a data stream (time series facts) in real-time. The result from the incremental changes confined to a sliding window is then merged with the materialized batch view from the serving layer to generate up-to-date analysis results. We adapted this architecture to use the batch processing components.

4   Problem Statement

After explaining the problem of number portability and Big Data approach facing to traditional solutions, we'll see how Big Data finds a solution for each of the problems found in number portability.

The **first problem** which refers to speed with which actually a customer can leak to competition, it's necessary to identify these customers immediately to avoid they go to the competition, and as solution, Big Data processes and delivers real-time prediction.

The **second problem** which refers to confidentiality of data, classical solutions work primarily with internal data of the company, ignoring public sources such as websites which contain data not structured but very valuable. Of course if the company decides to use public data to make predictions, must be sure that these are reliable. Big Data works with new data sources such as reviews on social networks. These reviews represent opinions of customers about mobile phone service and thanks to them we can predict who try to leave.

The **third problem** which refers to unstandardized and inconsistent data. Data used must be cleaned and standardized before starting a process of analysis; you can't work directly with transactional data, first a ETL process should be applied. As a solution, the Big Data approach processes data without the need for do a previous cleaning process.

Finally, **the fourth problem** which refers to the change of customer opinion about the service, since Big Data approach detects changes in data over time.

As we see, these four problems are solved by a Big Data approach seen in the Sect. 3, therefore, it's possible to build a Big Data architecture that implements a solution to prevent number portability.

5   A Big Data Architecture to Solve the Problem of Number Portability

After seeing the connection between the problems of classical solution and how Big Data and the 5 V offer a solution for each problem, we will explain the details of the architecture proposed in this paper.

5.1   Description of Solution

The solution proposed in this paper uses the new Big Data information sources, specifically social networks. The objective of the proposed solution is to obtain real-time comments left in the official Facebook fan pages of mobile phone operators, and analyze the sentiment expressed to predict if a customer try to realize a portability to competition.

5.2   Architecture Component

Figure 4 shows the architecture designed in this paper. It was designed with reference to Subsect. 3.1.

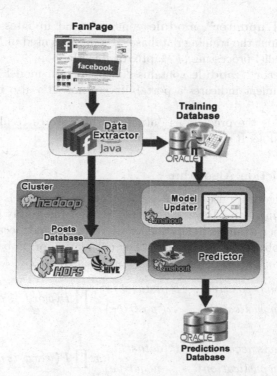

Fig. 4. Big Data Architecture for prediction of number portability.

This architecture consists of six components:

The **"data extractor" module** inspects the official Facebook page of the mobile phone operators and downloads the posts and comments made therein. The module extracts the data each hour and stores them in the "posts database". The data extractor module is also executed to generate the "training database" and to train the model. It's implemented in Java and uses the API OpenGraph to connect to Facebook.

The **"posts database" module** stores posts and comments of Facebook, saves the data on a distributed environment in HDFS of Hadoop. This module uses Hive to manage data with a syntax similar to SQL.

The **"training database"** is the component that stores posts and comments that will be used to train the prediction model. Each comment stored has a flag indicating if it's "negative" in case of comments that show intentions to leave, or "positive" in otherwise. The comments were manually labeled. To label a comment as "negative", it must meet at least one of following conditions: manifesting that a product or service from the competition is better, manifesting any complaint about a service or product of Claro Peru, manifesting directly that one will do a portability, or suggesting a portability to other customers. The way how each comment was labeled as positive or negative was suggested by business analysts of Claro Peru. This database is implemented as an Oracle relational database.

The **"model updater" module** generates and updates the prediction model using as input the training database. It's implemented in Mahout to take advantage of parallel processing of MapReduce in Hadoop.

The **"predictor" module** contains the prediction model and predicts if a customer's comment indicates a portability or not. It's also implemented in Mahout.

Finally, there is the predictions database which stores results obtained by the predictor module. It's implemented in Oracle.

5.3 The Prediction Algorithm

The prediction algorithm used in the architecture of Big Data is "Naive Bayes". In machine learning Naive Bayes classifier is part of the family of Bayes classifiers. This classifier assumes independence of variables and because of this the calculation of probabilities is simplified. For calculation of probabilities, we use formulas of [14] but adapted to our case (Formulas 1 and 2)

$$P \begin{pmatrix} publication\,is\,positive\,| \\ words\,in\,publication \end{pmatrix} = P \begin{pmatrix} dataset\,is \\ positive \end{pmatrix} x \prod P(word_i\,is\,positive) \quad (1)$$

$$P \begin{pmatrix} publication\,is\,negative\,| \\ words\,in\,publication \end{pmatrix} = P \begin{pmatrix} dataset\,is \\ negative \end{pmatrix} x \prod P(word_i\,is\,negative) \quad (2)$$

To improve the quality of the Naive Bayes classifier we used TF-IDF [15], which measures the relative importance of each word and makes that the least significant words (stopwords) would be ignored in the probability calculation. The Naive Bayes classifier algorithm was implemented in Java using the Mahout library.

5.4 Flow in Prediction Process

The prediction flow defined for our architecture is shown in Fig. 5.

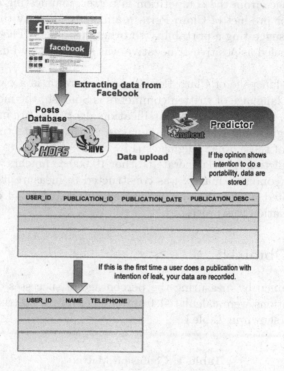

Fig. 5. Flow in prediction process.

First, comments are taken from the official Facebook fan page of mobile phone operator and are stored in the "posts database". This extraction is performed every hour looking for comments that have a date later than the last extraction performed. Then, the "predictor" module analyzes all comments obtained of the last extraction and for each one, predicts if it indicates portability or not. If the prediction indicates portability, the comment is stored in the predictions database, furthermore, if it's the first time that a user does a comment which indicates portability, the user data are also stored.

6 Tests

To analyze the behavior of our Big Data architecture, two types of tests were performed. The first of them measures the percentage of successes that the model has to predict the portability and the second measures the response speed of the cluster.

For first test, the prediction model was generated using Mahout and taking about 10,000 comments posted on the Facebook fan page of Claro Peru mobile operator. Each of these comments was labeled as "negative" if the opinion shows intention to do a portability, and "positive" otherwise. To label a comment as "negative", it must meet at least one of following conditions: manifesting that a product or service from the competition is better, manifesting any complaint about a service or product of Claro Peru, manifesting directly that one will do a portability, or suggesting a portability to other customers. The way how each comment was labeled as positive or negative was suggested by business analysts of Claro Peru.

As Facebook fan page of Claro Peru mobile operator is a page oriented to Peruvian public, language of written comments is Spanish, the most widespread language of Peru, however, the text classification algorithms are mostly designed to work with English and European languages, so this architecture can also be tested with any of these other languages [14].

With the prediction model generated, other 10,000 comments were loaded for validation and a confusion matrix was constructed to measure its effectiveness. With the generated model, the system was executed during 14 days, in which 336 GB of information was processed.

7   Results Obtained

The results obtained by measuring the percentage of successes of positive and negative publications were calculated from the confusion matrix generated by testing, which is shown in Table 1.

Table 1. Confusion Matrix

Real\Prediction	Positive	Negative
Positive	5,839	275
Negative	722	3,164

Concerning predictions made for positive comments, we got a total of 5,839 successes and a total of 722 errors, so success percentage for prediction of positive comments was 88.99%. Concerning predictions made for negative comments, we got a total of 3,164 successes and a total of 275 errors, so success percentage for prediction of negative comments was 92.00%. In general, for positive and negative comments, Big Data architecture proposed has obtained a success percentage of 90.03%.

8 Conclusions and Future Works

We obtained a state of the art accuracy of 90.03%. The architecture enables mobile phone company can identify in real time what customers intend to go to the competition, taking as source, public data left on social networks.

It was found that Big Data are related to issues raised in the Sect. 4 of the article and were resolved because the predictions were made quickly, it was worked with public data from Facebook, it was performed a process which the client was unaware of the analysis performed on his comments.

For the implementation of the architecture, Hadoop, Mahout and Hive technologies were used, other technologies such as Spark, Storm and Impala also help to improve the capabilities of an architecture in real time [16]. As future work, these Big Data tools could be implemented in the proposed architecture.

Finally, it's important to note that prediction of number portability plays an important role in industry of mobile telephony, because in order to reduce various costs associated with the loss of customers, it's imperative that mobile phone companies deploy predictive models to identify which customers will go to the competition.

References

1. Barrientos, F., Ríos, S.A.: Aplicación de Minería de Datos para Predecir Fuga de Clientes en la Industria de las Telecomunicaciones. In: Revista Ingeniería de Sistemas, vol. XXVII, pp. 73–107 (2013)
2. Organismo Supervisor de Inversion Privada en Telecomunicaciones (OSIPTEL): Estado de la portabilidad numérica en el primer trimestre del 2015, Perú (2015)
3. Kirui, C., Kirui, H., Hong, L., Cheruiyot, W.: Predicting customer churn in mobile telephony industry using probabilistic classifiers in data mining. Int. J. Comput. Sci. Issues 10(1), 165–172 (2013)
4. Pérez Villanueva, P.A.: Modelo de Predicción de Fuga de Cliente de Telefonía Móvil Post Pago. In: Memoria para optar al Título de Ingeniero Civil Industrial. Departamento de Ingenieria Industrial, Universidad de Chile, Chile (2014)
5. Kohavi, R.: a study of cross-validation and bootstrap for accuracy estimation and model selection. In: 14th International Joint Conference on Artificial Intelligence, vol. 2, pp. 1137–1143 (1995)
6. Morales, G.D.F., Bifet, A.: SAMOA: scalable advanced massive online analysis. J. Mach. Learn. Res. 16(16), 149–153 (2015)
7. Vasuki, M., Arthi, J., Kayalvizhi, K.: Decision making using sentiment analysis from twitter. Int. J. Innovative Res. Comput. Commun. Eng. 2(12), 71–77 (2014)
8. Kashyap, K., Deka, C., Rakshit, S.: A review on big data, hadoop and its impact on business. Int. J. Innovatite Res. Dev. 3(12), 78–82 (2014)
9. Gajjar, D.: Implementing the Naive Bayes classifier in Mahout. J. Emerg. Technol. Innovative Res. 1(6), 449–454 (2014)
10. Mukhopadhyay, D., Agrawal, C., Maru, D., Yedale, P., Gadekar, P.: Addressing namenode scalability issue in hadoop distributed file system using cache approach. In: 2014 International Conference on Information Technology, Bhubaneswar, India, pp. 321–326 (2014)

11. Kumar, R., Gupta, N., Charu, S., Bansal, S., Yadav, K.: Comparison of SQL with HiveQL. Int. J. Res. Technol. Stud. **1**(9), 28–30 (2014)
12. Banaei, S.M., Moghaddam, H.K.: Hadoop and its roles in modern image processing. Open J. Marine Sci. **4**(4), 239–245 (2014)
13. Dutta, K., Jayapal, M.: Big data analytics for real time systems. In: Big Data Analytics Seminar, pp. 1–13 (2015)
14. Mangal, S.B., Goyal, V.: Text news classification system using Nave Bayes classifier. Int. J. Eng. Sci. **3**, 209–213 (2014)
15. Mtafya, A.R., Huang, D., Uwamahoro, G.: On objective keywords extraction: Tf-Idf based forward words pruning algorithm for keywords extraction on youtube. Int. J. Multimedia Ubiquitous Eng. **9**(12), 97–106 (2014)
16. Barlow, M.: Real-Time Big Data Analytics: Emerging Architecture. O'Really Media (2013)

Social Networks of Teachers in Twitter

Hernán Gil Ramírez[✉] and Rosa María Guilleumas García[✉]

Universidad Tecnológica de Pereira, Carrera 27 #10-02, Pereira, Risaralda, Colombia
{hegil,roguiga}@utp.edu.co

Abstract. This research aimed at identifying the trends in the topics of interest of the tweets published by 43 expert professors in the field of ICT and education and the network of their followers and followed in Tweeter, as well as their relationship with the characteristics of that network. With this purpose, NodeXL was employed to import, directly and automatically, 185,517 tweets which gave origin to a network of connections of 49,229 nodes. Data analysis involved social network analysis, text extraction and text mining using NodeXL, Excel and T-Lab. The research hypothesis was that there is a direct correlation between the trends identified in the topics of interest and the characteristics of the network of connections that emerge from the imported tweets. Among the conclusions of the study we can highlight the following: (1) most of the trends identified from the analyzed tweets were related to education and educational technologies that could enhance teaching and learning processes; (2) the text mining procedure applied to the tweets revealed an interesting association between education and technologies; (3) and finally that the analysis of lemmas seems to be more promising than that of hashtags for detecting trends in the tweets.

1 Introduction

Currently, social networks in digital spaces are an important part of the life of a good number of people and institutions. Nevertheless, their study poses important challenges for researchers, since the huge volume of data circulating through them implies -for collection, processing, and analysis-, the use of specialized software, powerful equipment, complex analysis methods, and qualified people, items that are not always available in the small and middle-size educational institutions.

Though many users exchange through Twitter what Ferriter [1] calls "digital noise", this researcher claims that professionals in education have found ways to use Twitter to share resources and provide a quick support to colleagues with similar interests, turning this service into a valuable source of ideas to explore.

Twitter may be used for communication purposes, but also to share information and build, collectively, academic communities. This social network enables interaction with other people, access to their interests and identification of trends from the published messages.

J.A. Lossio-Ventura and H. Alatrista-Salas (Eds.): SIMBig 2015/2016, CCIS 656, pp. 133–145, 2017.
DOI: 10.1007/978-3-319-55209-5_11

2   Research Background

This work[1] takes as referents previous research on Twitter and the generation, exchange and propagation of information; it also considers works about the influence of users on this digital space. Shneiderman [2] explores the reasons for the success of social media like Facebook, Twitter, YouTube, Blogs, and the traditional discussion groups and concludes that it is due to the fact that they allow people to participate actively in local and global communities; the role of Twitter as a communication resource and information exchange tool during a crisis is tackled in Herverin and Lisl [3] research, and also in Chew and Eysenbach's [4] work.

Weng, Lim, Jiang and He [5] focus on the issue of the identification of the influential users of Twitter; Bakshy, Hofman, Mason and Watts [6] study the features and relative influence of Twitter's users. Regarding the propagation of information, our referents are Lerman and Ghosh [7], as well as the research carried out by Gómez, Leskovec, and Krause [8], where they state that the diffusion of information, and viral propagation are fundamental processes in the networks; we finally highlight the work done by Wu, Hofman, Mason and Watts [9], who stress the importance of understanding the channels through which information flows, in order to comprehend how it is transmitted.

3   Theoretical Considerations

Castells [10] thinks that the Internet is revolutionizing communication thanks to its horizontality, feature which permits users to create their own communication network and to express whatever they want, from citizen to citizen, generating a capacity of massive communication, not mediated by the traditional mass communication media. This communication networks are the basis of the "network society," a concept popularized by this author, who describes it as the social structure that characterizes the society of the early 21st century, a social structure constructed around (but not determined by) digital communication networks [11]. It is in the space and the time of the network society where the studied group of teachers constructs communication networks using Twitter, making out of it more than just a simple technology, a tool for communication, encounter, and assistance.

Castells defines a network as a set of interconnected nodes. The nodes may have more or less relevance for the network as a whole, so those of higher importance are called "centers" in some versions of the network theory. At any rate, any component of a network (including the "centers") is a node, and its function and meaning depend on the network programs and on its interaction with other nodes in it [11]. This author explains that the importance of the nodes in a network is higher or lower depending on how much important information they absorb and process efficiently, that is, it is determined by their capacity to contribute to the effectiveness of the network in the achievement of its programmed objectives (values and interests).

---

[1] This paper was developed within the framework of research project 4-14-5, *Dynamics of social networks of teachers in Twitter*, founded by the Univ. Tecnológica de Pereira (Colombia).

In this sense, we approach the study of the communication networks created by teachers from the connections they establish in Twitter. In this case, each user, and each web domain, hashtag, lemma, constitutes a node which establishes connections in the network under study, where it is evidenced that there are nodes with higher relevance than others. This is precisely what contributes to the understanding of the dynamics of these networks: what nodes are more important in the network, which are their contributions, and in what way they make up the structures of these relationships.

Social networks, as posed by Lévy [12], provide tools for human groups to join mental efforts so as to constitute intellects or collective imaginaries. This allows for connecting informatics to be part of a technical infrastructure of the collective brain of lively communities, which profit from social and cognitive individual potentialities for their mutual development. Lévy [12] describes collective intelligence as a type of intelligence that is disseminated everywhere, constantly enhanced, real-time coordinated and leading to an effective mobilization of competences and explains that the foundation and goal of the collective intelligences is mutual recognition and enrichment of people.

Concerning this point, we can sustain that networks like Twitter create the suitable space to integrate the intelligence of many people, located in different places around the world; an intelligence that is permanently updated, allowing people linked to the network to widen their horizons and possibilities to access information. Our intent in this research is, following Lévy's lead, to appraise the potential of Twitter as a space for interaction in the network of the teachers under study, and also to value the information they exchange and which can be accessed through this means, as a manifestation of collective intelligence.

4   Methodology

This research followed a quantitative approach with a trans-sectional, correlational, non-experimental design, which allowed for the establishment of the relation between the trends in the topics of interest detected and the structure of the network of connections that emerged from the tweets published by the selected group.

In order to select the group to be studied, we adapted the snowball sampling method. An initial group of seven (7) professors was intentionally identified and selected on the basis of their academic background related to the use of the ICTs in education, and their academic contributions via the Internet, in particular through Twitter. In a second phase, there was a follow-up of these seven professors' Twitter accounts, in order to identify other teachers who followed them or that they followed, and who, on the basis of their contributions in Twitter, their publications and academic output about the use of ICT in education, could be part of the studied group. This procedure was repeated once again until finally it was formed, in a not probabilistic way, a group of 43 teachers.

Of the selected group, 65% were University professors, 23% primary and secondary teachers and 12% belonged to other type of institutions (non-formal, virtual tutors and advisors). Concerning their nationalities, 84% were from Spain, 7% from Argentina, 5% from Colombia, 2% from Mexico and 2% from Venezuela.

Using NodeXL we imported from Twitter, 185,517 tweets published by the network of connections of the 43 selected teachers between February the 4th and June the 6th, 2014.

As data collection instruments, we used NodeXL templates (which include not only the tweets but also the information of the edges, as well as that of the nodes). From the imported data rose a network of connections made up 49,229 nodes and 98,494 edges.

These nodes were located in 128 countries. 88.3% of them were concentrated in 10 countries, among them, Spain, Argentina, The United States, Colombia, and Mexico. About one third of the nodes registered in their profile professions related with education.

In order to identify the trends in the topics of interest in the published tweets and their relationship with the features of the network from which they emerged, we made a graphic representation of the network and calculated its metrics, using NodeXL. Likewise, we identified the trends in the topics of interest by analyzing the imported tweets to quantify the frequencies of appearance of the hashtags and by applying text mining to the content of the tweets. We also identified the trends in the web domains and established the correlation among the frequencies of the topics of interest detected as trends and the metrics of the network, using multivariate analysis, and Pearson's correlation coefficient. For data analysis we used the programs NodeXL, Excel, T-Lab and Statgraphics.

5   Analysis and Data Interpretation

For data analysis and interpretation, we examined the features of the network of connections of the 43 teachers selected. Besides, based on the tweets published by the mentioned network, we identified the trends in the topics of interest and studied their correlation with the values obtained in the two previous steps.

5.1   Features of the Communication Network

We used NodeXL to make the graph of the network of connections as well as to calculate its metrics.

Taking a look at Fig. 1 (below) with its 49,229 nodes and 98,494 edges, it is evident that, given their location, not all the nodes have the same importance in the network. A representative group of nodes, located in the center, are the most connected; a significant amount, the least connected, are displaced outwards, and a couple of them, though connected to each other, are disconnected from the network.

Figure 2 (below) corresponds to the same network after the application of a filter based on the nodes' Betweenness Centrality index and it shows only those with a value higher than 1 for that index. This action produced a reduction of the network to 8,725 nodes (a 17.7% of the total) and facilitated to note, more clearly, the set of nodes that occupied the center, while in the periphery, in opaque tones, there can be seen the remaining nodes, those out of the established filter.

Fig. 1. Communication network emerging from the imported tweets.

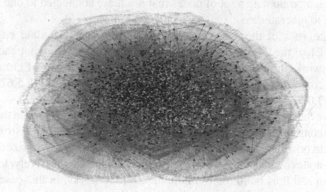

Fig. 2. Communication network emerging from the imported tweets, filtered.

Thus, we can see the configuration of a network, that as Castells sustains [11], is made up of interconnected nodes; some, the so-called centers, of greater importance for the network, and others, less important, depending on their capacity to access information and process it efficiently; that is, on account of their capacity to contribute to the achievement of the objectives of the network itself.

The process of analysis implied, likewise, the calculation of the graph's metrics, as a basis for the quantitative measurement of the indices associated to the nodes and their edges. The graph was directed. The relation of reciprocity of the edges was of 0.27. The In-Degree ranged between 0 and 3,439, the Out-Degree between 0 and 1,789 and the Betweenness Centrality index between 0 and 354,805,308.32.

Of the 49,229 nodes analyzed, the 10 nodes with a higher In-Degree, Out-Degree, and Betweenness Centrality, belonged to the initial group of 43 teachers selected. This shows that, in addition to a relatively high level of edges between the nodes of the network, the initial group of 43 teachers selected, from which the network of connections emerged, had a significant weight within the network, both for the amount of nodes connected to them as for the amount of nodes to which they were connected and therefore for their intermediation potential in the network. This is particularly important in a scenario where just a few nodes had high degrees of intermediation.

The 49,229 nodes of the network were organized in 24 groups of diverse sizes, according to the number of nodes in them. There was a high amount of edges inside each group, as well as among the different groups. For instance, group 1 had 8,220 nodes (16.7% of the total) and 174,397 edges. At the other end in size and edges were group 23 (with 525 nodes, 1.1% of the total and 586 connections) and group 24 (disconnected from the network, with just 2 nodes and a single edge between them).

Regarding the making up of the groups, we want to state that within a network of connections it is difficult to establish groups as well as their precise borders since the nodes can be involved in different relations and belong to more than one group.

In this research, the clusters were conformed with NodeXL, using the Clauset-Newman-Moore algorithm for clusters, that automatically identifies the groups from the network structure, placing the densely connected nodes in separated groups; that is, conforming each group with a set of nodes that are more connected to one another than what they are to other nodes.

On average, each of the 24 groups had 2,051 nodes, 2,939 inner edges and was connected to 21 of the 24 existing groups through 1,164 edges, what shows a highly connected network. In this respect, it is worth noting the existence of groups that were rather highly connected to other groups, as for example, group 1 with 5,678 edges, and group 2 with 3,076.

We believe that the existing communication among the nodes, inside the conformed groups and among them, facilitates the access to information and its distribution inside the studied network, thanks to what Castells [10] calls the process of horizontality, which allows all the nodes connected to the network to communicate massively, to share whatever they wish and thus build their communication networks, in this case through the use of Twitter.

As a summary, we can affirm that the network studied was decentralized, though not in the classic sense of the term since some nodes were connected to one or more central nodes, which in turn were often connected to several nodes, central or not, making the structure of this network more complex and robust, in such a way that if one of the central nodes were to disappear, this would not cause the disconnection of a great amount of nodes or the disappearance of the network.

The study of the tweets exchanged in the studied network showed that, within it, the identified trends (hashtags, lemmas and web domains) were the origin of other networks.

5.2   Identification of Tendencies of the Topics of Interest to Be Published

The web domains referenced in the tweets, as well as the hashtags and slogans more used, led to the identification of the trends in the topics of interest to be published in the studied network.

Tendencies identified from the hashtags referenced in the tweets. Of the 185,517 imported tweets, 31.5% (58,349) included hashtags. The total of referenced hashtags was 88,798, out of which 29,590 were unique hashtags. We identified the hasthtags referenced in the tweets and calculated their frequency of appearance. The 10 hashtags with a higher referencing frequency (0.03% of the total) were used 6% of the times, while the remaining 29,590 (99.97%) appeared the remaining 94% of the times. The first place was for the hashtag #education, followed by #ABPmooc_intef and #elearning, #tic, #edtech, #eduPLEmooc, among others.

The ten hashtags with a higher frequency of use in the tweets could be grouped around three main topics: education (8 hashtags), politics (1 hashtag), and technology (1 hashtag). The predominance of the hashtags related to the topic of education could seem obvious in a network initially composed by teachers; however, we should remember that the 43 initially selected teachers were the seed of a network that was enlarged to include 49,299 nodes; this suggests that the said 43 teachers followed and were followed either mainly by teachers, or by people interested in and concerned about education.

This piece of data may show some degree of homophily in the studied network of connections, since despite the fact that Twitter users are not forced to correspond to their followers (directed network) and most of the links are not corresponded, the users tend, however, to connect to others exhibiting interests and activities similar to their own [13]. This situation also matches Wu, Hofman, Mason and Watts's findings [9] who highlight the significant homophily found in their research.

Network of tendencies identified from the 10 most referenced hashtags. The tendencies identified from the 10 most referenced hashtags enabled the conformation of a network of connections between the nodes referencing the hashtags (source node) and the hashtags which were being referenced (target node).

Most of the connections were grouped around a specific hashtags. There are very few cases in which a node used more than one hashtag. However, as an example of this case, we can mention #eduPLEmooc y #ABPmooc_intef, which set up some connections with the same users.

Figure 3 (below) was the result of the application of a filter based on the Betweenness Centrality index of the nodes. It shows the 154 nodes (8.2% of the total) with a higher than the average value of this index. This process allowed the visualization of those nodes with greater force of intermediation in the network, located in the central part of the graphic. It also let us observe that most of them, about 91.8%, had a low or no force of intermediation at all. These nodes, represented with opaque tones, were located in the periphery of the graphic according to the decreasing value of the index, a value that reached 0 for 1,533 nodes, that is, for the 81.3%.

As we can observe from these metrics, there was an important number of nodes which could be considered as "lurkers", since they do not contribute much to the network; they are mainly silent participants.

The In-Degree index in this network ranged between 0 and 335, the Out-Degree between 0 and a 6; and the Betweenness Centrality between 0 and 1,453,757.65. Although a hashtag can receive many entries (as in the case of #education, with an In-Degree of 335, or #TheKingAbdicates, with 237), these are generated by many nodes. For this reason, we can assert that the tendencies detected are actually a product of the

Fig. 3. Network of connections of the 10 most referenced hashtags, filtered.

individual contributions of an important number of network nodes, what evidences the materialization of Lévy's collective intelligence.

Within the network of connections of the 10 hashtags with a higher frequency of use in the tweets, 21 groups were conformed. On average, each group connected only to 2 other groups, and there were even some groups that were not connected to any other. It is remarkable that the groups with a larger number of nodes connected to a greater amount of groups. One example of this is Group 1, which having 271 nodes, was connected to 5 groups. In contrast, the groups with a lower number of nodes showed a tendency to not setting up connections to any group. This was the case of group 21, which having 2 nodes, did not connect to any group.

Trends identified in the lemmas of the tweets. In order to advance in the identification of the topics of interest in the tweets published by the network of connections of the selected group of teachers, we resorted to text mining. The analysis of the content of the tweets was done with T-Lab, using the automatic lemmatization (word grouping) and the selection of key words.

Starting from the 185,517 imported tweets, the corpus of analyzed data was made up of 175,122 elementary contexts (EC), 179,374 words, 162,072 themes, and 2,574,255 occurrences. The program automatically selected the 500 words with the highest levels of occurrence in the corpus, out of which the non-meaningful terms were manually deleted later (articles, preposition, etc.) giving a remainder of 310 items. For text segmentation (elementary contexts), we used the paragraph, which in this case was equivalent to a tweet. For the selection of key words we employed the method of occurrences.

Lemmas associated with education, such as *education, educational, learning or course* stood out in frequency of citation in the tweets as shown in Table 1. The lemma *education* had already been identified also as one of the 10 most referenced hashtags.

Table 1. Lemmas and Elementary Contexts (EC)

N°	Lemmas	EC	N°	Lemmas	EC
1	Education	4,024	6	Course	2,238
2	New	2,543	7	Follow	2,201
3	Educational	2,415	8	Blog	2,143
4	Social	2,404	9	Stories	2,117
5	Learn	2,303	10	Life	2,063

Analysis of co-occurrences/word associations. The co-occurrence is the number of times (frequency) that a lexical unity (LU) appears in the corpus or within the elementary contexts (EC), in this case in the tweets. The function word association was used to detect which words, in the elementary contexts, co-occurred with the lemma *education*.

Education, found in 4,024 of the 175,122 elementary contexts (EC) analyzed, was associated to a group of lemmas, considered as relatively close, among them *ict* and *technology*. Their relationship was confirmed by the higher values of the index of association: *ICT*, 0.166: *technology*, 0.166, since the closer the association between two lemmas, the higher the coefficient.

In addition, the lemma *ict* appeared in 1,577 elementary contexts, and the lemmas *education* and *ict* were referenced together in 419 elementary contexts. As we can observe in Table 1, there was evidence of the prevalence of lemmas associated with education, as well as of the close association between them, in the elementary contexts analyzed.

Tendencies of the web domains identified in the tweets. Out of the 185,517 imported tweets, 59.4% included references to web domains. Using Excel, 113,361 domains were identified, out of which 18,448 were unique web domains. In order to detect the tendencies in the domains, we calculated their frequency of reference and located the 10 with the highest levels of reference. It is worth noting the great amount of references accumulated by these 10 domains, since making up just for a 0.05% of the amount of unique domains found in the tweets, they were referenced in the 25.4% of the occasions.

Among the 10 most cited web domains were blog sites (blogspot, 1st position), sites for video publishing (Youtube, 2nd position); social networks (Facebook, 3rd position; Instagram, 6th position; LinkedIn, 7th position; Foursquare, 9th position); online newspapers and journals (Paper.li, 5th position; eldiario.es, 10th position); content curation sites (Scoop.it, 4th position).

It should be highlighted that most of the referenced domains (4 out of 10) were social network applications. Likewise, we must point out the importance of the blogs for the studied network, since besides the tweets that included mentions to blogs of blogspot, there was also a considerable amount of domains making reference to other blogs, like in the case of blogs.elpais, blog.educalab, blog.tiching, blog.fernandotrujullo and blogthinkbig.

This list of web domains in general and blogs in particular permits the visualization of tendencies in the use of the web, and may help teachers approach the best possibilities to explore them and integrate them in their teaching practices.

Network of the tendencies of the web domains identified in the tweets. The 10 more cited domains in the tweets allowed shaping a network of connections between these 10 web domains (target nodes) and the users referencing them (source node). This new network was made up of 10,900 nodes and 28,745 connections (7,319 unique connections and 21,426 duplicated connections).

To facilitate the analysis and interpretation of the graph, we applied a filter based on the Betweenness Centrality Index of the nodes, allowing the visualization of the nodes with a higher power of intermediation in the network, and therefore, with a greater significance in so far as the flow of information.

Figure 4 shows the 1,397 nodes (about 12.8% of the total) with a higher than the average value. These were the small group of nodes located in the center of the graph. These nodes may be crucial in the flow of information, since they lied in the paths between other nodes in the network and therefore provided a link between them.

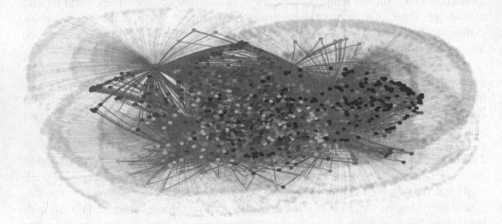

Fig. 4. Network of connections of the 10 most referenced domains

Toward the periphery, in opaque tones, we can see the remaining nodes, the ones left outside by the applied filter. The great majority of them had a low Betweenness Centrality index, which reached 0 for 9,295 nodes, that is, for the 85.3%. These values reflect a distribution of Pareto, in which a small number of nodes (about 13%) displayed the higher values of Betweenness Centrality, while a great number of nodes (87%) showed relatively low values in this index.

The In-Degree Index of this network had a minimum value of 0 and a maximum of 3,367; The Out-Degree presented a minimum of 0 and a maximum of 6; the Betweenness Centrality showed a minimum of 0 and a maximum of 56,149,284.89. These metrics evidenced a higher maximum value of In-Degree than of Out-Degree, what indicates that though a web domain may have been referenced many times (as the in the case of Youtube, with an In-Degree of 3,367), these references were done by many nodes. In other words, we can assert that the detected tendencies were actually a product of Lévy's collective intelligence, and not of reduced groups of nodes that fostered a particular interest.

Nine groups were configured inside the connection network of the 10 domains with the highest frequencies of appearance in the tweets. Group 1, despite being the most numerous, did not connect to any other groups, though other groups did connect to it. On average, each group established connections with five other groups; the average amount of nodes by group was 1,211 and that of the unique connections, 704.

We must highlight that the groups with a lower amount of nodes established connections with a greater amount of groups, to the point that groups 8 and 9 were connected to 8 of the 9 groups configured, while the groups with a greater amount of nodes –groups 1 and 2- were connected to less groups (0 and 1 group respectively). This could mean that a great number of the network nodes posted tweets referencing a particular web domain, while a minority of them, referenced in their tweets a greater variety of web domains.

6   Correlation Between Tendencies and Metrics

In order to correlate the six (6) variables associated with the network of connections under study, we applied a multivariate analysis, relating pairs of variables of the metrics with the frequencies of the identified trends. The variables of the metrics were: In-Degree, Out-Degree, and Betweenness Centrality. The variables of the tendencies were: web domains (URL), hashtags, and lemmas.

As shown in Table 2, in most of the relations between pairs of variables of the metrics and the tendencies of the topics of interest, we found a direct correlation, though weak.

Table 2. Correlations

	In-Degree	Out-Degree	Betweenness Centrality
URL	0.162	**0.172**	0.146
Hashtag	0.046	0.054	0.045
Lemma	0.196	**0.201**	**0.183**

The highest correlation was observed between lemmas and metrics, and the lowest between hashtags and metrics. In the first case, the highest correlation occurs between lemmas and out-degree, followed by lemmas and in-degree.

As evidenced in the results, the higher it is a node's outdegree, it is more probable that the lemmas and URL that they publish may appear among the identified tendencies.

This finding attests to the importance of considering the URL and lemmas published in the tweets of those users with the highest outdegree values in order to determine the predominant tendencies associated to the use of ICT in education.

7   Conclusion

The methodological procedure used in this research allowed us to create a wide network of users interested in education starting from an initial group of 43 teachers.

Although the nodes of the initial group registered high values in the network metrics, their influence in the identified trends was low.

Most of the trends identified from the analyzed tweets were related to education and educational technologies that could enhance teaching and learning processes, as for instance, blogs, social networks as platforms for sharing documents and other resources, online journals and curation tools.

It stands out the association between education and technologies found through the text mining procedure applied to the tweets.

The importance of blogs as a trend was confirmed by its appearance among the web domains with the highest frequency of references in the tweets.

The direct correlation found particularly between the metrics of the network and the trends in the lemmas found in the analysis of the tweets, allows to conclude the importance of analyzing with particular attention the tweets published by users with a high out-degree since they seemed to influence more the trends that arise from the studied network.

The analysis of lemmas seems to be more promising than that of hashtags for detecting trends in the tweets.

Since nearly 6 of each 10 tweets included a reference to a web domain, it would be interesting to be able to explore in a greater detail, what users are actually referencing through those web domains.

The results of this research and their usefulness for identifying trends in the topics of interest of educational professionals suggest the potential usefulness of continuing exploring the possibilities of social networks and the analysis of big data in the shaping academic communities.

References

1. Ferriter, W.M.: Why teachers should try Twitter. Educ. Leadersh. **67**(5), 73–74 (2010)
2. Shneiderman, B.: Technology-mediated social participation: the next 25 years of HCI challenges. In: Jacko, Julie, A. (ed.) HCI 2011. LNCS, vol. 6761, pp. 3–14. Springer, Heidelberg (2011). doi:10.1007/978-3-642-21602-2_1
3. Heverin, T., Lisl, Z.: Microblogging for crisis communication: examination of Twitter use in response to a 2009 violent crisis in the Seattle-Tacoma, Washington Area. In: Proceedings of the 7th International ISCRAM Conference (2010)
4. Chew, C., Eysenbach,G.: Pandemics in the age of Twitter: content analysis of Tweets during the 2009 H1N1 outbreak. PLoS ONE **5**(11) (2010)
5. Weng, J., Lim, E., Jiang, J., He, Q.: TwitterRank: finding topic-sensitive influential twitterers. In: Proceedings of the Third ACM International Conference on Web Search & Data Mining, pp. 261–270 (2010)
6. Bakshy, E., Hofman, J.M., Mason, W.A., Watts, D.J.: Everyone's an influencer: quantifying influence on Twitter. In: Proceedings of the fourth ACM international conference on Web search and data mining (WSDM 2011), pp. 65–74 (2011)
7. Lerman, K., Ghosh, R.: Information contagion: an empirical study of the spread of news on digg and Twitter social networks. In: Proceedings of 4th International Conference on Weblogs and Social Media (ICWSM) (2010)

8. Gómez, M., Leskovec, J., Krause, A.: Inferring networks of diffusion and influence. In: Proceedings of the 16th ACM SIGKDD International Conference on Knowledge Discovery and Data Mining, pp. 1019–1028 (2010)
9. Wu, S., Hofman, J.M., Mason, J.M., Watts, D.J.: Who says what to whom on Twitter. In: Proceedings of the 20th International Conference on World Wide Web, pp. 705–714 (2011)
10. Castells, M.: Internet y la sociedad red (2001)
11. Castells, M.: Comunicación y poder. Alianza, España (2009)
12. Lévy, P.: Inteligencia colectiva. Por una antropología del ciberespacio. OPS/OMS (2004)
13. Kwak, H., Lee, C., Park, H., Moon, S.: What is Twitter, a social network or a news media? In: Proceedings of the 19th International Conference on World Wide Web, pp. 591–600 (2010)

Author Index

Printed in the United States
By Bookmasters